Lecture Notes in Computer Science 11648

More information about this series at http://www.springer.com/series/7407

Chris Thachuk · Yan Liu (Eds.)

DNA Computing and Molecular Programming

25th International Conference, DNA 25
Seattle, WA, USA, August 5–9, 2019
Proceedings

 Springer

Editors
Chris Thachuk
California Institute of Technology
Pasadena, CA, USA

Yan Liu
Arizona State University
Tempe, AZ, USA

ISSN 0302-9743   ISSN 1611-3349   (electronic)
Lecture Notes in Computer Science
ISBN 978-3-030-26806-0   ISBN 978-3-030-26807-7   (eBook)
https://doi.org/10.1007/978-3-030-26807-7

LNCS Sublibrary: SL1 – Theoretical Computer Science and General Issues

This Springer imprint is published by the registered company Springer Nature Switzerland AG
The registered company address is: Gewerbestrasse 11, 6330 Cham, Switzerland

Preface

This volume contains the papers presented at DNA 25: the 25th International Conference on DNA Computing and Molecular Programming. The conference was held at the University of Washington in Seattle, Washington, USA during August 5–9, 2019, and was organized under the auspices of the International Society for Nanoscale Science, Computation, and Engineering (ISNSCE). The DNA conference series aims to draw together researchers from the fields of mathematics, computer science, physics, chemistry, biology, and nanotechnology to address the analysis, design, and synthesis of information-based molecular systems.

Papers and presentations were sought in all areas that relate to biomolecular computing, including, but not restricted to: algorithms and models for computation on biomolecular systems; computational processes in vitro and in vivo; molecular switches, gates, devices, and circuits; molecular folding and self-assembly of nanostructures; analysis and theoretical models of laboratory techniques; molecular motors and molecular robotics; information storage; studies of fault-tolerance and error correction; software tools for analysis, simulation, and design; synthetic biology and in vitro evolution; and applications in engineering, physics, chemistry, biology, and medicine. Authors who wished to orally present their work were asked to select one of two submission tracks: Track A (full paper) or Track B (one-page abstract with supplementary document). Track B is primarily for authors submitting experimental results who plan to submit to a journal rather than publish in the conference proceedings. We received 39 submissions for oral presentations: 19 submissions to Track A and 20 submissions to Track B. Each submission was reviewed by at least four reviewers, with an average of five reviewers per paper. The Program Committee accepted 12 papers for Track A and 13 papers for Track B. Additionally, the Program Committee reviewed and accepted 61 submissions to Track C (poster) and selected seven for short oral presentation. This volume contains the papers accepted for Track A.

To commemorate the 25th anniversary of the conference, Microsoft Research sponsored a special session on Thursday that included a panel discussion and reception. The conference concluded with a Molecular Technology Day on Friday consisting of invited and contributed talks.

We express our sincere appreciation to our invited speakers, Leonard Adleman, David Baker, Michael Elowitz, Dana Randall, Anne Shiu, and Erik Winfree, our invited Molecular Technology Day speakers, Kate Adamal, Andy Ellington, Jocelyn Kishi, and Jeff Nivala, and our panelists. We especially thank all of the authors who contributed papers to these proceedings, and who presented papers and posters during the conference. Last but not least, the editors thank the members of the Program Committee and the additional invited reviewers for their hard work in reviewing the papers and providing constructive comments to authors.

August 2019

Chris Thachuk
Yan Liu

Organization

Steering Committee

Anne Condon (Chair)	University of British Columbia, Canada
Luca Cardelli	University of Oxford, UK
Masami Hagiya	University of Tokyo, Japan
Natasha Jonoska	University of South Florida, USA
Lila Kari	University of Waterloo, Canada
Chengde Mao	Purdue University, USA
Satoshi Murata	Tohoku University, Japan
John Reif	Duke University, USA
Grzegorz Rozenberg	University of Leiden, The Netherlands
Nadrian Seeman	New York University, USA
Friedrich Simmel	Technical University of Munich, Germany
Andrew Turberfield	University of Oxford, UK
Hao Yan	Arizona State University, USA
Erik Winfree	California Institute of Technology, USA

Program Committee

Chris Thachuk (Co-chair)	California Institute of Technology, USA
Yan Liu (Co-chair)	Arizona State University, USA
Mark Bathe	Massachusetts Institute of Technology, USA
Robert Brijder	Hasselt University, Belgium
Yuan-Jyue Chen	Microsoft Research, Redmond, USA
Anne Condon	University of British Columbia, Canada
Mingjie Dai	Harvard University, USA
David Doty	University of California, Davis, USA
Andre Estevez-Torres	CNRS, France
Elisa Franco	University of California, Los Angeles, USA
Cody Geary	Aarhus University, Denmark
Anthony J. Genot	CNRS, France and University of Tokyo, Japan
Manoj Gopalkrishnan	India Institute of Technology, Bombay, India
Elton Graugnard	Boise State University, USA
Masami Hagiya	University of Tokyo, Japan
Rizal Hariadi	Arizona State University, USA
Natasha Jonoska	University of South Florida, USA
Lila Kari	University of Waterloo, Canada
Yonggang Ke	Emory University, USA
Matthew R. Lakin	University of New Mexico, USA
Chenxiang Lin	Yale University, USA
Satoshi Murata	Tohoku University, Japan

Pekka Orponen	Aalto University, Finland
Thomas Ouldridge	Imperial College London, UK
Matthew Patitz	University of Arkansas, USA
Lulu Qian	California Institute of Technology, USA
John Reif	Duke University, USA
Dominic Scalise	Johns Hopkins University, USA
Joseph Schaeffer	Autodesk Research, USA
William Shih	Harvard University, USA
David Soloveichik	University of Texas at Austin, USA
Darko Stefanovic	University of New Mexico, USA
Andrew Turberfield	University of Oxford, UK
Bryan Wei	Tsinghua University, China
Shelley Wickham	University of Sydney, Australia
Erik Winfree	California Institute of Technology, USA
Damien Woods	Maynooth University, Ireland

Additional Reviewers

Aubert-Kato, Nathanael	Mohammed, Abdulmelik
Bui, Hieu	Schweller, Robert
Durand-Lose, Jérôme	Severson, Eric
Eftekhari, Mahsa	Shah, Shalin
Fu, Dan	Song, Tianqi
Hader, Daniel	Stewart, Jaimie
Haley, David	Subramanian, Hari
Hendricks, Jacob	Summers, Scott
Imai, Katsunobu	Sun, Wei
Jacobson, Bruna	Viswa Virinchi, Muppirala
Kawamata, Ibuki	Wang, Wen
Linder, Johannes	Zadegan, Reza M.
Meunier, Pierre-Étienne	

Local Organizing Committee for DNA 25

Georg Seelig (Chair)	University of Washington, USA
Luis Ceze	University of Washington, USA
Karin Strauss	Microsoft Research, Redmond, USA and University of Washington, USA
Max Willsey	University of Washington, USA

Sponsors

International Society of Nanoscale Science, Computing, and Engineering
Microsoft Research (USA)
National Science Foundation (USA)
University of Washington

Contents

Chemical Reaction Networks
and Stochastic Local Search

Erik Winfree[(⊠)]

California Institute of Technology, Pasadena, CA, USA
winfree@caltech.edu

Abstract. Stochastic local search can be an effective method for solving a wide variety of optimization and constraint satisfaction problems. Here I show that some stochastic local search algorithms map naturally to stochastic chemical reaction networks. This connection highlights new ways in which stochasticity in chemical reaction networks can be used for search and thus for finding solutions to problems. The central example is a chemical reaction network construction for solving Boolean formula satisfiability problems. Using an efficient general-purpose stochastic chemical reaction network simulator, I show that direct simulation of the networks proposed here can be *more* efficient, in wall-clock time, than a somewhat outdated but industrial-strength commercial satisfiability solver. While not of use for practical computing, the constructions emphasize that exploiting the stochasticity inherent in chemical reaction network dynamics is not inherently inefficient – and indeed I propose that stochastic local search could be an important aspect of biological computation and should be exploited when engineering future artificial cells.

1 Introduction

Can a cell solve Sudoku? While few would take seriously the prospect of whether an individual *E. coli* could beat the puzzle page of the *Daily Mail*, this question of principle has significant implications. Sudoku, when generalized, is a particularly relatable example of an NP-complete problem, and it has been effectively used to illustrate methods to solve constraint satisfaction problems [1,2] as well as to explore neural computing architectures underlying natural intelligence [3,4]. So our real question is whether 1 cubic micron of biochemistry could efficiently implement the kinds of algorithms necessary to solve difficult problems like Sudoku, and if so, how? An answer to this more general question could be valuable for engineering a future artificial cell that makes the most of its limited computing resources to get by in the world; it may conceivably also provide new perspectives on the principles exploited by naturally evolved cells to solve the problems they encounter during their lives.

We will use the P $\overset{?}{=}$ NP question to frame our discussion [5]. Informally, P is the class of problems that are *solvable* in polynomial time (with respect to

© Springer Nature Switzerland AG 2019
C. Thachuk and Y. Liu (Eds.): DNA 25, LNCS 11648, pp. 1–20, 2019.
https://doi.org/10.1007/978-3-030-26807-7_1

the size of the problem instance). A canonical problem in P is the CIRCUITE-
VALUATION problem, where a problem instance specifies a feedforward Boolean
circuit and values for each input wire, and the solution is what the circuit should
output on the given input. To solve such an instance requires simply evaluating
each gate in order, which requires only polynomial (roughly linear) time. Informally, NP is the class of problems whose solutions are *verifiable* in polynomial
time (where a solution now is the answer accompanied by a polynomial-sized
certificate that explains why). A canonical problem in NP is the CIRCUITSAT-
ISFIABILITY problem, where a problem instance specifies a feedforward Boolean
circuit and values for each output wire, and the solution is whether there exist
input values for which the circuit produces the specified output; those input
values constitute a certificate that can be verified in polynomial (roughly linear)
time. CIRCUITSATISFIABILITY is effectively the inverse problem for CIRCUITE-
VALUATION. Problems in NP always can be solved within exponential time by
brute-force enumerating all possible certificates, and verifying them one by one
– if any certificate checks out, then an answer has been found. The $P \overset{?}{=} NP$
question essentially asks whether there are better algorithms for NP problems
that do something much more clever and thus are guaranteed to find solutions
in polynomial time – if this is possible for all NP problems, then P = NP. But
most computer scientists think the answer is "no": while clever algorithms may
reduce the form of the exponential, or may provide polynomial time solutions
for a subset of cases, nonetheless for the worst-case hard problem instances of
NP problems, no algorithm can guarantee a polynomial time to find a solution.

What interests us here is *not* whether P = NP or not, but rather the fact
that studying this question has revealed how fundamental and natural the class
NP is. Presuming that P ≠ NP, then there are problems (many problems!) that
are often enough very hard to solve, yet recognizing that a solution has been
found is relatively easy. Problems with this character have substantial impact
in the real world. For example, in cryptography, decoding a message is hard
without a key, but easy with the key – and this makes internet commerce feasible.
But problems with this character can be found far beyond computer science
and technology. For example, in academia, a professor managing a large group
is feasible only because solving research problems is hard and time-consuming
(the graduate student's job) but recognizing that the research problem has been
solved (the professor's job) is dramatically easier and less time-consuming. In
fact, the professor is able to steer the research group in the direction she wants
just by wisely posing a set of problems to be tackled, waiting for the students
to solve them (by cleverness, by luck, by standard techniques, by unexpected
genius, by brute force, it doesn't matter), and then verifying that each solution
is genuine. More generally, hierarchical organization in science, business, and
society relies on the distinction between the difficulty of doing a task and the
ease of verifying that it has been done properly. The design of organizations
that are effective at accomplishing their goals in turn relies on aptly defining the
subtasks and the criteria for success.

Let's call this *programming by recognition*: rather than specifying all the details of an algorithm for how to *find* a solution, the programmer just provides a simpler specification for how to *recognize* a solution. This is in fact a central idea for constraint logic programming languages, such as PROLOG and CLP, where the programmer concerns himself with what (easy to check) constraints define the solution that is being sought, and standardized methods are used to find the solution [6]. Programs in such languages often don't have guarantees on run times, but when done carefully, they can often find solutions effectively.

From this perspective, CIRCUITSATISFIABILITY can be viewed as a very low-level constraint logic programming language, and along with more powerful generalizations such as satisfiability modulo theories, Boolean satisfiability solvers are now used routinely in industry [7–10]. While carefully-crafted deterministic algorithms remain the dominant general-purpose methods for solving Boolean constraint problems, for many years a broad class of hard problems were best solved by surprisingly simple stochastic algorithms that perform biased random walks through the space of potential solutions [11–14]. This observation positions stochastic local search as a viable engine to power systems designed using the programming by recognition paradigm.

There is ample evidence for programming by recognition in biology, suggesting that it provides evolutionary advantages as a system architecture. A classic example occurs during Eukaryotic cell division, when it becomes necessary to move one copy of each chromosome to polar opposite sides of the cell. Since the interior of a cell can be chaotic, this poses the problem of how to find each chromosome, prior to pushing them into place. Nature's solution is to grow microtubules bridging pairs of chromosomes, using a "search and capture" approach [15] whereby microtubules grow in all directions but are stabilized only when they recognize their proper target. The search involves stochastic exploration and energetically expensive mechanisms such as "dynamic instability" that alternate fast growth with fast depolymerization, which has been shown to be a more effective algorithm than a passive and energetically-neutral random walk of polymer length [16]. Compared to a hypothetical deterministic molecular algorithm for solving the same problem ("take a left at the endoplasmic reticulum..."), the stochastic "search and capture" algorithm presumably has the advantages that it is simpler to encode genetically, more robust, and thus more easily evolvable [17, 18]. Moving to a higher level of biological organization, a second classic example is learning, which often involves a biased random walk through parameter space that recognizes when good performance has been achieved [19, 20]. As before, there are advantages: organisms that learn can have smaller genomes, are more robust, and can evolve faster than organisms with hard-coded circuitry [21]. Indeed, evolution itself exhibits some characteristics of programming by recognition – if survival is the ultimate form of recognition.

Our interest here is to explore programming by recognition as a paradigm for engineering circuits within artificial cells that exploit the natural stochasticity of molecular interactions to provide efficient solutions to hard problems that the cell might encounter. For simplicity, we ignore the geometric factors present in the motivating example of cell division and instead focus on well-mixed chemical reaction networks involving a finite number of species in a finite volume. This choice is attractive not only for its mathematical simplicity, but also because it can be considered a programming language for engineering molecular systems using dynamic DNA nanotechnology [22–25].

2   Stochastic Chemical Reaction Networks

We will use the standard model for formal chemical reaction networks (CRNs) with discrete stochastic semantics for mass-action kinetics [26]. A CRN is specified by a finite set of species, e.g. $\{A, B, C, \ldots, X, Y, Z\}$ and a finite set of reactions, e.g. $\{A + B \xrightarrow{k_1} B, X + A \xrightarrow{k_2} C, Y \xrightarrow{k_3} A + C, A \xrightarrow{k_4} B\}$, where k_i are the stochastic rate constants with respect to an implicitly given reaction volume. The state of the CRN at time t is given by the discrete counts of each species, which we will refer to (with a mild abuse of notation) as A_t, B_t, etc. The propensity ρ_i of a reaction i at time t gives the instantaneous rate of firing for that reaction, and may be calculated as the product of k_i and the current counts of each reactant (if all reaction stoichiometries are 1), e.g. $\rho_1(t) = k_1 A_t B_t$ in the example above. The total propensity $\rho = \sum_i \rho_i$ is the sum over all reactions. This implicitly defines a continuous time Markov chain (CTMC) with an infinite state space (although, depending on the CRN and initial state, only a finite subset may be reachable). In the standard stochastic simulation algorithm (SSA), the time at which the next reaction occurs is chosen from an exponential distribution with mean $1/\rho$, and at this time, the reaction that occurs is reaction i with probability ρ_i/ρ. Simulation continues until a fixed time is exceeded, or a state is reached where $\rho = 0$.

Since our central thesis is that stochastic chemical kinetics allow surprisingly effective problem-solving, it is worth reviewing how stochastic chemical kinetics (appropriate for a finite volume containing discrete counts of each species) differs from the more familiar deterministic chemical kinetics (which models a hypothetical well-mixed but infinite volume where real-valued concentrations are the appropriate measure and ordinary differential equations provide the appropriate dynamics). I will offer six perspectives.

Noisy behavior. The classical view is that stochastic CRNs are just noisy versions of deterministic CRNs. There are senses in which this is true. For non-explosive CRNs [27], if the volume V of a stochastic CRN is scaled up while keeping the initial concentrations of each species constant, then for chosen time t, the concentrations in the stochastic model approach the concentrations in the deterministic model [28]. This is a Central Limit Theorem result, that is, X_t approaches a Gaussian with coefficient of variation shrinking with V. But it is not a uniform convergence: to ensure that the stochastic CRN's concentrations are likely within a certain percentage of the deterministic CRN, V may have to grow with t (depending on the specific CRN).

Extinction and reachability. For a fixed volume, stochastic CRN behavior can differ qualitatively from deterministic behavior. Most notably, species (or reactions) may become extinct in the stochastic CRN despite being active forever in the deterministic CRN. This occurs in the classical predator-prey oscillator, $\{R \rightarrow 2R, F + R \rightarrow 2F, F \rightarrow \emptyset\}$. More generally, stochastic CRNs can be limited by discrete state-space reachability constraints, reflecting deep connections to computability theory [29–31] that are not present in the continuous deterministic state space.

Perfect Boolean values. The hard distinction between *none* and *some* that is inherent to stochastic CRNs can make deterministic computation easier than in deterministic CRNs. For example, implementing feedforward Boolean logic circuit computations is straightforward with stochastic CRNs [22], where signals on wires can be represented by counts of 0 or 1 for specific species. But when signals are represented as high or low real-valued concentrations in deterministic CRNs, error-correcting non-linearities are needed to perform signal restoration [32].

Computing with counts. The count for a single chemical species can store the value of an arbitrary non-negative integer, giving stochastic CRNs the ability to perform uniform computation – in the sense that a single computing machine can process arbitrarily large input instances [33–36]. Turing-universal computation is possible with vanishingly small probability of error.

Computing with distributions. Probabilistic behavior and probabilistic reasoning is an adaptive response for living in a world where partial information is available. Unlike deterministic CRNs, where probabilities must be represented as concentrations [37,38], stochastic CRNs have the potential to directly represent probabilities via probabilities [39–41]. Although stochastic CRNs that satisfy detailed balance can only produce equilibrium distributions that are products of Poisson distributions [42], constraints of state-space reachability can shape marginals to approximate arbitrary distributions, as can non-equilibrium steady-states in CRNs that do not satisfy detailed balance [43].

That was only five. The sixth perspective – **that stochastic CRNs inherently perform stochastic local search** – is developed in the rest of this paper.

3 Evaluating and Satisfying Circuits

We begin by reviewing how stochastic CRNs can efficiently solve CIRCUITE-VALUATION problems, using methods similar those in work cited above. Given a feedforward Boolean circuit that uses 2-input unbounded fan-out gates, our goal is to construct a stochastic CRN that, presented with molecules representing any input to the circuit, will produce molecules representing the output of the circuit, and then all reactions will become extinct. Specifically, for a circuit c with N inputs and M gates, the CRN CIRCUITEVALUATIONCRN[c] will employ $2(N + M)$ species and $4M$ reactions, as illustrated in Fig. 1(A) and as follows:

(A)

N inputs, M gates

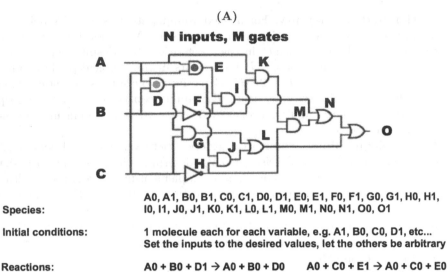

Species:	A0, A1, B0, B1, C0, C1, D0, D1, E0, E1, F0, F1, G0, G1, H0, H1, I0, I1, J0, J1, K0, K1, L0, L1, M0, M1, N0, N1, O0, O1

Initial conditions: 1 molecule each for each variable, e.g. A1, B0, C0, D1, etc...
Set the inputs to the desired values, let the others be arbitrary

Reactions:

A0 + B0 + D1 → A0 + B0 + D0   A0 + C0 + E1 → A0 + C0 + E0
A0 + B1 + D1 → A0 + B1 + D0   A0 + C1 + E1 → A0 + C1 + E0
A1 + B0 + D1 → A1 + B0 + D0   A1 + C0 + E1 → A1 + C0 + E0
A1 + B1 + D0 → A1 + B1 + D1   A1 + C1 + E0 → A1 + C1 + E1

...   ...

(B)

A0 + B0 + D1 → A0 + B0 + D0   A0 + B0 + D1 → A0 + B1 + D1   A0 + B0 + D1 → A1 + B0 + D1
A0 + B1 + D1 → A0 + B1 + D0   A0 + B1 + D1 → A0 + B0 + D1   A0 + B1 + D1 → A1 + B1 + D1
A1 + B0 + D1 → A1 + B0 + D0   A1 + B0 + D1 → A1 + B1 + D1   A1 + B0 + D1 → A0 + B0 + D1
A1 + B1 + D0 → A1 + B1 + D1   A1 + B1 + D0 → A1 + B0 + D0   A1 + B1 + D0 → A0 + B1 + D0

...   O1 → O0   ...

Fig. 1. (A) Evaluating feedforward Boolean circuits with stochastic CRNs.
Red reactions correspond to the gate with the red dot; blue reactions correspond to
the gate with the blue dot. **(B) Solving Circuit-SAT with stochastic CRNs.** Red
reactions correspond to the gate with the red dot. Clamping reactions set the output
to the desired value. In this example, $N = 3$ and $M = 12$. (Color figure online)

- Each wire in the circuit is named, and for each wire X we use two species, $X0$ and $X1$.
- Initially, and for all time thereafter, there will be one molecule per wire, i.e. $X0_t + X1_t = 1$.
- For each gate $Z = g(X, Y)$ where g is a Boolean function, the four reactions are

$$X0 + Y0 + Z(1 - v) \rightarrow X0 + Y0 + Z(v) \qquad \text{with } v = g(0,0)$$
$$X0 + Y1 + Z(1 - v) \rightarrow X0 + Y1 + Z(v) \qquad \text{with } v = g(0,1)$$
$$X1 + Y0 + Z(1 - v) \rightarrow X1 + Y0 + Z(v) \qquad \text{with } v = g(1,0)$$
$$X1 + Y1 + Z(1 - v) \rightarrow X1 + Y1 + Z(v) \qquad \text{with } v = g(1,1)$$

where $Z(0) \equiv Z0$, $Z(1) \equiv Z1$, and all rate constants are identical, say k.

In other words, for each input case for the gate, there are reactions that catalytically convert an incorrect output species into a correct output species. It will be convenient to associate full states of the CRN, in which there are exactly one molecule for each wire, with assignments of Boolean values to each wire, in the obvious way. We can generalize this to partial assignments, so that e.g. for circuit input x, WIRES$[x]$ refers to any full state of the CRN such that the species representing input wires have the specified values, while other species have arbitrary counts consistent with the one-molecule-per wire constraint. Similarly, for circuit output y, WIRES$[y]$ refers to any full state of the CRN such that the output species have the specified values, and other species are unconstrained.

Theorem 1 (Feedforward circuit evaluation). *For feedforward Boolean circuit c, input x, and stochastic CRN* CIRCUITEVALUATIONCRN$[c]$, *any initial state* WIRES$[x]$ *will evolve to a state* WIRES$[c(x)]$ *at which point all reactions will be extinct.*

Proof. Since c is feedforward, we can order the gates such that their inputs are always either prior gates or inputs to the circuit. Because each reaction is catalytic in the gate inputs, once all wires prior to a gate are correct, they will never change again. That gate will become correct with expected time $1/k$. Thus, all M wires will be correct within expected time $O(M/k)$.   □

Feedforward circuits compute from input to output in polynomial time. To reverse the information flow from output to input, thus solving a circuit satisfaction problem, requires some guesswork or brute-force enumeration, and (if P $\neq$ NP) requires more than polynomial time, e.g. exponential time. Surprisingly, the stochastic CRN for solving this problem is not significantly larger, in terms of the number of reactions.

The idea remains that each CRN reaction will detect a violation of circuit correctness, such that if all reactions go extinct, then the circuit represents a valid solution to the problem. There are two new ingredients. First, unlike how we set the inputs for feedforward evaluation using the initial state, here we will allow arbitrary initial state, but treat the output species having the wrong value as a violation of circuit correctness. Therefore, we introduce clamping reactions that detect if the output is wrong, and if so, fix it. E.g. if output wire Y should be 0, we include a reaction $Y1 \rightarrow Y0$. Second, when a gate's output is inconsistent with its input, we no longer know whether the problem is that the output is incorrect, or the input is incorrect – so we also include reactions that detect the gate's inconsistency and change one of the inputs, instead of changing the output. These reactions are illustrated in Fig. 1(B). Which reaction fires first is random, according to stochastic CRN semantics. Altogether, for circuit c and output y, this construction results in a circuit CIRCUITSATCRN$[c, y]$ with $2(N + M)$ species and $12M + L$ reactions, where there are L output wires. Note that the circuit no longer needs to be feedforward, and the clamped target values do not need to be output wires; our comments below can be trivially extended to this generalization.

Theorem 2 (Solving circuit satisfiability). *For satisfiable Boolean circuit* *c, output y, and stochastic CRN* CIRCUITSATCRN[*c, y*], *any initial state will evolve to a state* WIRES[*x, y*] *where y = c(x), at which point all reactions will be extinct.*

Proof. Since the set of possible local violations of correctness correspond exactly to the set of reaction reactants, it follows that if all reactions go extinct, then a solution has been found. Similarly, if the current CRN state does not correspond to a solution, then at least one reaction in the CRN will be poised to fire. What remains to be shown is that, from any initial state, a state representing a valid solution can be reached. Let s be a CRN state corresponding to a valid solution, and let s' be the current state, which is not a valid solution. Suppose s' deviates from s in n wires. If s' is not valid because of an incorrect output wire, then we can take a reaction that corrects the output wire, leading to a state s'' that has $n-1$ deviations from s. If s' is not valid because of an incorrect gate, then either an input or the output of that gate deviates from s. One of the 12 reactions for that gate corrects this deviation, leading to a state s'' that has $n-1$ deviations from s. Either way, if s'' is not itself a valid solution, then we recursively show that there is a sequence of reactions that leads either to s or to another valid solution. □

Thus, when CIRCUITSATCRN[*c, y*] is started in any one-molecule-per-wire state, it will eventually find a solution x if one exists. However, if no solution exists – i.e. the circuit is *not* satisfiable – then the CRN will continue exploring the state space forever. And even when a solution does exist, we have no useful bound on how long the CRN will take to find it[1].

4   Formula Satisfiability

The computer science community has converged not on circuit satisfiability problems, but on formula satisfiability problems as the standard for Boolean logic constraint solvers. Circuit satisfiability problems can be translated to formula satisfiability problems with little overhead; in fact, even when limited to clauses with no more than three literals each – i.e. 3-SAT – the problem is still NP-complete. Such simple building blocks facilitate analysis and constructions. Figure 2(A) illustrates how to construct stochastic CRN FORMULASATCRN[*f*] that solves 3-SAT formula f using the same principles as for circuit satisfiability, but now using just $2N$ species and $3M$ reactions, where f has N variables and M clauses. A sample run is shown in Fig. 2(B). This CRN has similar merits and demerits as the circuit satisfiability CRN: in theory, it is guaranteed to eventually find a solution if one exists; in practice, you're not likely to have the patience to watch it try to solve a hard problem with 100 variables.

---

[1] The sequence of reactions identified in the proof will have a positive probability of occurring next, specifically, at least $(12M + L)^{-(N+M)}$. This provides an exponential bound on the expected time to halt, to wit, less than $(12M + L)^{N+M}/k + (N+M)/k$. But is that useful?

(A)

Given: a boolean formula	(A or L or N) and (A or !K or !N) and (!J or !P or !R) and (Q or E or I) and (D or !A or !K) and (B or T or C) and (O or I or !Q) and (G or !O or !Q) and (K or D or F) and (T or !M or !B) and (M or !S or !B) and (R or C or !T) and (R or !P or !C) and (S or F or !D) and (L or I or !E) and (I or !L or !E) and (M or S or !F) and (H or !O or !G) and (P or !J or !H) and (J or !G or !H) and (L or E or !I) and (E or !L or !I) and (L or !E or !I) and (!L or !E or !I)
Construct: a chemical reaction network that stops when it finds a solution to the formula	
Species:	A0, A1, B0, B1, C0, C1, D0, D1, E0, E1, F0, F1, G0, G1, H0, H1, I0, I1, J0, J1, K0, K1, L0, L1, M0, M1, N0, N1, O0, O1, P0, P1, Q0, Q1, R0, R1, S0, S1, T0, T1
Initial conditions:	1 molecule each for each variable, e.g. A1, B0, C0, D1, etc...

Reactions:

A0 + L0 + N0 → A1 + L0 + N0	A0 + K1 + N1 → A1 + K1 + N1
A0 + L0 + N0 → A0 + L1 + N0	A0 + K1 + N1 → A0 + K0 + N1
A0 + L0 + N0 → A0 + L0 + N1	A0 + K1 + N1 → A0 + K1 + N0
J1 + P1 + R1 → J0 + P1 + R1	
J1 + P1 + R1 → J1 + P0 + R1	...
J1 + P1 + R1 → J1 + P1 + R0	

(B)

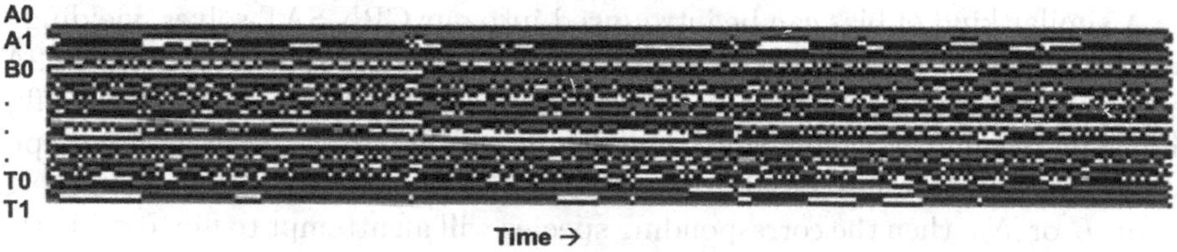

Fig. 2. (A) Solving 3-SAT problems with stochastic CRNs. Red reactions correspond to the red clause; blue reactions correspond to the blue clause; green reaction correspond to the green clause. The $(N = 20, M = 24)$ formula illustrated here was once solved on a DNA computer [44], although variables have been renamed. (B) Space-time history of a 3-SAT run. Species are arranged vertically; each column of 40 pixels corresponds to the state of the CRN before or after each reaction fires; black if the species count is 0, grey if the species count is 1 but that disagrees with the solution eventually found, white if the species count is 1 and it agrees with the solution eventually found. At the end of time, the CRN has gone extinct. (Color figure online)

Why is our formula satisfiability CRN so ineffective? One intuition is that when a clause is not satisfied, there are three possible ways to try to fix it – flip one of the three variables – but some of those actions will cause other clauses to break. The CRN makes no distinctions, so in a situation where there is one way to make things better, and two ways to make things worse, it is more likely to make things worse. Thus, as the CRN performs a stochastic local search of the state space, the projection onto the number of conflicted clauses does a random walk that hovers around some average number, making occasional excursions toward more or toward fewer.

Similar problems have been encountered in conventional SAT-solvers based on stochastic local search, and effective – though heuristic – countermeasures have been devised [11–14]. The basic idea is to bias the choice of clause and variable to flip so as to favor moves that reduce the number of conflicts. A particularly simple and effective incarnation of this idea is WalkSAT [12]. The core algorithm is just this:

1. Start with a random assignment of values to variables.
2. At random choose an unsatisfied clause.
3. Flip a variable from that clause, thus making it satisfied, such that the fewest other clauses become unsatisfied.
4. With some low probability, flip a random variable.
5. If the formula is not satisfied, go back to step 2.

Variants of this algorithm dominated the "random" category in the international SAT competitions for over a decade, until 2014 [9]. (For formulas generated randomly with a certain ratio of clauses to variables, $\alpha = M/N$, there is a critical value of $\alpha \approx 4.26$ below which problems are likely to be satisfiable and "easy" to solve, and above which problems are likely to be unsatisfiable; near the threshold, problems become quite hard [14, 45, 46].)

A similar kind of bias can be introduced into our CRN SAT-solver, yielding a new construction WALKSATCRN[f], which now has $4N$ species and $4M + 2N$ reactions. What seems simplest to implement in a CRN is to reject attempts to flip bits that, if changed, would cause many clauses to become unsatisfied. Our specific construction is illustrated in Fig. 3(A). If there is an unsatisfied clause, e.g. (A or L or N), then the corresponding species will all attempt to flip, e.g. via the reaction $A0 + L0 + N0 \rightarrow tryA1 + tryL1 + tryN1$. While they are trying to flip, these variables cannot be involved in additional conflict detection events, thus somewhat limiting the number of variables that simultaneously try to flip. However, at the beginning of the run, a "wound area" of variables trying to flip will quickly emerge, possibly until all clauses either have some flipping variable or are already satisfied. Now there is a competition for "healing" the wound: at a slow rate, variables trying to flip will successfully solidify their choice via reactions such as $tryA1 \rightarrow A1$; while at a faster rate flips will be rejected if solidifying would have introduced a conflict, e.g. via the rejection reaction $D0 + tryA1 + K1 \rightarrow D0 + A0 + K1$ associated with clause (D or !A or !K). Thus, the healing wound will result in changed values preferentially for variables that introduce no new conflicts, or few of them – with the preference being more strongly against changes that introduce more conflicts, because the rate of rejecting an attempted flip will be proportional to the number of reactions trying to reject it. Nonetheless, occasionally changes will increase the number of conflicts, since which reaction occurs next is probabilistic in the SSA.

Whereas a simple argument sufficed to show that CIRCUITSATCRN[c, y] and FORMULASATCRN[f] halt with a valid solution if and only if their problem is satisfiable, the fact that WALKSATCRN[f] attempts to flip three variables simultaneously has so far confounded my attempts to prove its correctness[2]. Moreover, we have no theoretical guarantees for the effectiveness of the stochastic local search bias for finding solutions quickly.

Therefore, without further analysis and sticking with the first fast-to-slow reaction rate ratio that seemed to work OK, we evaluated the effectiveness of WALKSATCRN[f] on random satisfiable formulas near the critical threshold. Formulas with 100 variables were reliably solved in just a few seconds, while formulas with 500 variables were reliably solved in under an hour ($10^{3.56}$ seconds) and typically much faster. In no case did we encounter a satisfiable formula that the CRN failed solve. Several comments are in order. First, how fast the CRN runs, in wall-clock time, depends on what CRN simulator you use. Second, it would be useful to have a point of comparison for how hard it is to solve the random formula instances.

For the comparison, we used Mathematica's built-in command **SatisfiableQ**, which makes use of the MiniSAT 1.4 library [47]. MiniSAT is a deterministic algorithm that can both solve satisfiable problems as well as declare problems to be unsatisfiable. MiniSAT and its variants have been perennial winners in the international SAT competition, although in recent competitions the improved MiniSAT 2.2 has merely been the baseline that the winning algorithms soundly surpass [9].

Regarding the simulator, I used a general-purpose Mathematica-based CRN simulator originally developed by Soloveichik [48] that I extended to be more efficient on large CRNs by using data structures similar to those in prior work [49–51]. Specifically, the simulator pre-computes a fixed binary tree of reactions in which all reaction propensities are stored, along with a list Δ_i, for each reaction i, indicating which other reactions' propensities will be changed when reaction i fires – i.e. those whose reactants have their counts changed by the given reaction. Thus, for a CRN with R reactions, each step of the SSA involves $\lg R$ choices through the binary tree to navigate to the selected reaction i, followed by recalculation of the propensities of only the reactions in Δ_i, followed by $|\Delta_i| \lg R$ hierarchical updates of propensities within the binary tree. This avoids much of the redundant calculations in naive implementations of SSA. For further speed, the inner loop of the SSA uses Mathematica's **Compile** function, which effectively compiles to C. Thus, the CRN simulator we use here is reasonably fast, but has no optimizations that are specific to the SAT-solving CRN constructions – it is a general-purpose CRN simulator.

Figure 3(B) compares the wall-clock time for running the CRN simulator on WALKSATCRN[f] for random hard formulas f (selected by **SatisfiableQ** to be satisfiable) versus the time taken by **SatisfiableQ** itself to solve the same problem. For a given problem size (i.e. dot color, labeled by N for $M = 4.2N$),

---

[2] A straightforward adaptation of the previous argument works for a closely related CRN that is identical to WALKSATCRN[f] except that species $tryX0$ and $tryX1$ are conflated as $tryX$ for each variable X. This CRN should work similarly, as the main difference is merely that a variable being flipped now might spontaneously revert. But it is not the CRN that I simulated.

(A)

Given: a boolean formula

(A or L or N) and (A or !K or !N) and (!J or !P or !R) and
(Q or E or I) and (D or !A or !K) and (B or T or C) and
(O or I or !Q) and (G or !O or !Q) and (K or D or F) and
(T or !M or !B) and (M or !S or !B) and (R or C or !T) and
(R or !P or !C) and (S or F or !D) and (L or I or !E) and
(I or !L or !E) and (M or S or !F) and (H or !O or !G) and
(P or !J or !H) and (J or !G or !H) and (L or E or !I) and
(E or !L or !I) and (L or !E or !I) and (!L or !E or !I)

Construct: a chemical reaction
network that stops when it finds
a solution to the formula

Species:

A0, A1, B0, B1, ..., T0, T1
tryA0, tryA1, tryB0, tryB1, ..., tryT0, tryT1

Initial conditions:

1 molecule each for each variable, e.g. A1, B0, C0, D1, etc...

Reactions:

fast

A0 + L0 + N0 → tryA1 + tryL1 + tryN1   1. Try to fix a problem.
tryA0 + L0 + N0 → A1 + L0 + N0   2. Prevent another
A0 + tryL0 + N0 → A0 + L1 + N0   clause's reaction
A0 + L0 + tryN0 → A0 + L0 + N1   making a change that
                                        would cause a problem.
slow   ...   3. If no one complains,
        tryA0 → A0   make it so.
        ...

(B)

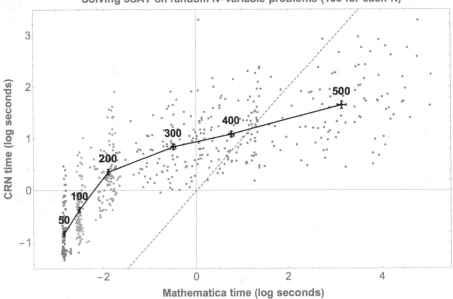

Solving 3SAT on random N–variable problems (100 for each N)

Fig. 3. (A) A WalkSAT-inspired CRN for solving 3-SAT problems with
stochastic CRNs. Fast reactions have rate constant 1.0, while slow reactions have
rate constant 0.1. (B) Wall-clock time comparison for solving random 3-SAT
problems. Each dot represents a different random 3-SAT formula with $M = 4.2N$
and the indicated number of variables. Times are reported by their logarithm base 10.
Blue dashed line indicates where Mathematica's **SatisfiableQ** takes the same time as
simulating the CRN. A 2.6 GHz Macbook Pro (2013) was used. (Color figure online)

there is considerable scatter in the times taken to solve different instances. The vertical scatter (the CRN time) is due not only to the variability in the problem instance, but also due to variability in the CRN's stochastic local search process – solving the same large instance multiple times reveals a timing standard deviation nearly equal to the mean. (In contrast, Mathematica's solver exhibited only 3% variation when re-solving the same problem, which was presumably due to background processes in the operating system.) The CRN simulator clearly has a substantial overhead: for small ($N = 50$) problems, the average CRN time is about a third of a second, while Mathematica solves most problems in little more than a millisecond. Despite this, the CRN becomes more efficient for larger problems, eventually out-performing Mathematica on the largest ($N = 500$) instances. Here, the CRN is taking on average about 5 minutes, while Mathematica is averaging at over two hours. For this N, the CRN has roughly 10,000 reactions.

Today's best SAT-solvers can beat random 3-SAT problems with $\alpha = 4.267$ and $N > 5000$ in under an hour [9]. I have not tried it, but I don't presume the WALKSATCRN[f] simulation would be anywhere close to being competitive. The take-home message is not that simulating a CRN is a good way to solve SAT problems – I don't think it is – but rather that stochastic local search comes fairly naturally to CRNs, and they are not incredibly bad at it. For a given SAT problem, the size of the CRN is linear in the number of variables and clauses (specifically, $4M+2N$) and the volume required to store one molecule per wire is just $O(N)$. (Note, however, that for a DNA strand displacement implementation, one DNA fuel complex would be required for each time a reaction fires, entailing a ridiculously unrealistic overhead since solving SAT problems of the scale shown in Fig. 3(B) involves millions or billions of reaction events. For fuel-efficient SAT-solving CRNs, see the beautiful papers by Thachuk and colleagues [52–54]; their CRNs effectively implement a brute-force enumeration method, but using reversible reactions within polynomial volume.)

5   Recognizing and Generating Patterns

We are tempted to think of SAT-solving by stochastic local search as a general-purpose mechanism for programming by recognition in molecular systems. The hard SAT instances may not represent the most interesting or useful cases; the value might be in the flexibility and robustness of the computing paradigm.

Consider the problem of making inferences about complex environmental signals. Formulas, circuits, or most generally, networks of relations [55] can be used to represent knowledge about a domain in terms of constraints. In the absence of information from the environment, there may be an enormous number of possible solutions; perhaps we only need one. When additional information is available, in the form of certain variables being True or False, all we need are clamping reactions (such as $A1 \rightarrow A0$ if the environment dictates that A is False), and the stochastic local search will be forced to find a solution that also satisfies the environmental constraints. Depending on the structure of the network and which variables are clamped, this may correspond to evaluating a circuit in the feedforward direction (easy!) or solving a circuit satisfiability problem (hard!) or something in between. It is a general form for potentially omnidirectional computing in CRNs by clamping. (There are striking similarities to the Chemical Boltzmann Machine [40], which allows for similarly omnidirectional inferences to be made by clamping species representing known information, but in that work the inference is probabilistic and computes conditional probabilities.)

A classic paradigm for omnidirectional computation is the associative recall of memories in the Hopfield model [56]. The task here is to store a finite set of "memories" (i.e. binary vectors) into the weights of a neural network, such that when the network is initialized with a test vector, the dynamics of neural updates will bring the system to the known memory that is "closest" (in some sense) and halt there. For example, suppose the memories are 17×17 binary images of Alan Turing, George Boole, Ludwig Boltzmann, Rosalind Franklin, Ada Lovelace, and Warren McCulloch. If the network can "see" only the top half of an image – i.e. the neurons corresponding to those pixels are clamped to the correct values for one of the memories, while the rest of the neurons continue to be updated – then it will reconstruct the rest of the image by "associative recall". It can do the same if it sees any sufficiently large subset of pixels. If the pixels that it sees are partially corrupted, then when the clamping is released, the errors will be corrected and the stored memory established – at least most of the time, with some caveats, and if not too many memories are stored.

A very similar associative memory task can be accomplished by a Boolean SAT solver performing stochastic local search. To "store" the memories, one needs to construct a formula (or circuit) that has several valid solutions – one for each memory. Now, if enough variables are clamped such that only one memory is compatible with the clamped variables, the SAT solver is guaranteed to find that unique solution, thus performing associative recall. If, instead, the variables are initialized (but not clamped) to a pattern similar to one of the memorized patterns, then the stochastic local search is likely to reach that solution first. We demonstrate this idea by constructing a formula whose only solutions are exactly those shown in Fig. 4(A), using the Exclusion Network approach [55]. Specifically, we randomly choose n triples of variables, for each triple make a subformula that allows only the combinations that occur in the set of target memories, and create a formula that is the conjunction of all n subformulas. This formula is guaranteed to have the target memories as valid solutions, and as n increases, it excludes more and more alternative solutions (if there are not too many target memories). After algebraically converting to conjunctive normal form for 3-SAT, we can build the CRN SAT-solver.

The constraints imposed by a SAT formula may be used to define not only pattern recognition processes, but also pattern generation processes. To illustrate the use of SAT solving for a morphogenesis application, we consider a case where, rather than having a unique solution or a small number of chosen solutions as in the associative memory discussed above, the SAT constraints define a combinatorial set of solutions with meaningful properties in common. Figure 4(B) shows several solutions of a SAT formula that imposes only local restrictions that

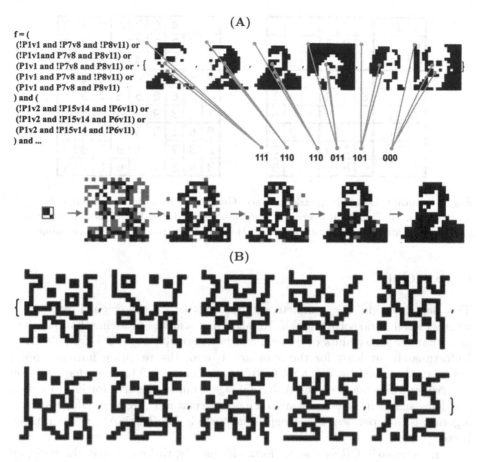

Fig. 4. (A) The WalkSAT-inspired CRN performing as an associative memory. Top: The principle for constructing a Boolean formula encoding the memories. Bottom: Clamping a 9-pixel portion of Boltzmann leads to recall of Boltzmann. Grey pixels are trying to flip. **(B) The WalkSAT-inspired CRN maintaining a membrane-like structure.** The Boolean formula only enforces that solutions have black in the corners and all interior cells are either white or have exactly two immediate neighbors that are black. Ten representative solutions are shown.

amount to insisting that black pixels form uninterrupted lines that connect to the corners. Thus, these patterns are fixed points for the SAT-solving CRN, but if disturbed by external perturbations – or additional internal constraints – the CRN will perform stochastic local search to re-establish the connectivity of the membrane-like patterns. Here we see that stochastic local search simultaneously forms the pattern and provides a self-healing capability.

These examples are intended to highlight the connection between stochastic local search SAT solvers and more biological notions of memory, inference, morphogenesis, self-healing, homeostasis, and robustness – how a biological organism self-regulates to restore and maintain critical structures, concentrations, and functions [57].

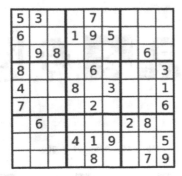

Fig. 5. Sudoku. Left: The puzzle instance. Given the black digits, the remaining cells must be filled in such that every 3×3 square has all digits $1 \ldots 9$, as does every row and every column. Right: The solution to this puzzle, in red. (Color figure online)

6 Sudoku

This paper wouldn't be complete without a return to the original question of whether a cell could solve Sudoku. To take a first stab at answering that question, note that the constraints of Sudoku can be expressed as a SAT formula [1,2]. Unfortunately, at least for the approach I used, the resulting formulas could not be easily solved by WALKSATCRN[f] simulations. This is perhaps not too surprising; while the classical WalkSAT algorithm is effective for hard random problems and many easy problems, on most "structured" problems, deterministic algorithms that perform clever depth-first search, like MiniSAT, perform much better [14].

Can stochastic CRNs also perform efficient depth-first search? It would be easier – and nearly as effective – to repeatedly take random dives through the search tree to a leaf, rather than to deterministically traverse each node exactly once, which would require extra bookkeeping. Such a stochastic depth-first search would bear some similarity to dynamic instability in microtubules' search for chromosomes [15,16]: the microtubule quickly grows in a random direction, making random choices as it goes; if it is unsuccessful finding a chromosome, it eventually suffers a "catastophe" and rapidly contracts, then starts over... until it finds a chromosome.

Based on this vision, we can construct a stochastic CRN to solve an arbitrary Sudoku puzzle (Fig. 5). There are 9^3 species whose presence indicates that a specific digit may possibly be in a specific cell; another 9^3 that indicate it is definitely not; another 9^2 that indicate that a cell's value is known (i.e. only one digit may be there); more that indicate it is unknown; and some additional bookkeeping species, including *Forward* and *Reverse*. When *Forward* is present, a set of reactions quickly enforce logical constraints among the species; more slowly, a cell with few options will spontaneously choose one, thus descending the search tree. If a contradiction is noted, a reaction will convert *Forward* to *Reverse*, and the logic bookkeeping will dissolve... to be rebuilt when *Reverse* switches to

Forward again. The CRN has roughly 35,000 reactions, and solves all available Sudoku problems in under half an hour, including the hardest ones on Gordon Royle's list [58].

The number of reactions in the Sudoku CRN is within an order of magnitude of existing whole-cell models of bacteria [59]. So maybe we could conceive of a bacterial-sized artificial cell that implemented that many reactions. However, successfully solving hard Sudoku puzzles involves many dives into the search tree and many millions or billions of reaction events. Even assuming rate constants on the order of 1 per second, that would take many days or years. *E. coli* reproduces in 20 min. So no, it seems that cells are unlikely to be successful solving Sudoku.

7 Discussion

Nonetheless, we may have learned something during this somewhat stochastic exploration of ideas. Foremost in my mind is that the stochasticity inherent in CRNs provides a natural engine for stochastic local search and thus programming by recognition – the hallmark algorithmic architecture defining NP problems. The architecture is robust and flexible, pivoting seamlessly from efficient solution of easy problems to relatively efficient solution of hard problems, naturally accommodating memory, inference, self-healing, and homeostasis with respect to constraints.

Software. Mathematica packages and notebooks for the CRN simulator, SAT-solving CRN constructions, and Sudoku solver can be found on the author's website [60].

Acknowledgements. This work was supported in part by National Science Foundation (NSF) grant 1317694 – The Molecular Programming Project. Thanks to Matt Cook, David Soloveichik, Chris Thachuk, William Poole, Lulu Qian, Grzegorz Rozenberg, Moshe Vardi, Tony Rojko, and Henry Lester for stimulating questions, comments, and encouragement.

References

1. Simonis, H.: Sudoku as a constraint problem. In: Proceedings of the 4th International Workshop on Modelling and Reformulating Constraint Satisfaction Problems, vol. 12, pp. 13–27 (2005)
2. Lynce, I., Ouaknine, J.: Sudoku as a SAT problem. In: Proceedings of the 9th International Symposium on Artificial Intelligence and Mathematics (AIMATH) (2006)
3. Hopfield, J.J.: Searching for memories, Sudoku, implicit check bits, and the iterative use of not-always-correct rapid neural computation. Neural Comput. **20**, 1119–1164 (2008)
4. Jonke, Z., Habenschuss, S., Maass, W.: Solving constraint satisfaction problems with networks of spiking neurons. Front. Neurosci. **10**, 118 (2016)
5. Garey, M.R., Johnson, D.S.: Computers and Intractability. W. H. Freeman and Company, New York (1979)

6. Jaffar, J., Maher, M.J.: Constraint logic programming: a survey. J. Logic Program. **19**, 503–581 (1994)
7. Vardi, M.Y.: Boolean satisfiability theory and engineering. Commun. ACM **57**, 5 (2014)
8. Järvisalo, M., Le Berre, D., Roussel, O., Simon, L.: The international SAT solver competitions. AI Mag. **33**, 89–92 (2012)
9. Balyo, T., Heule, M.J.H., Jarvisalo, M.: SAT competition 2016: recent developments. In: Thirty-First AAAI Conference on Artificial Intelligence (2017)
10. de Moura, L., Bjørner, N.: Z3: an efficient SMT solver. In: Ramakrishnan, C.R., Rehof, J. (eds.) TACAS 2008. LNCS, vol. 4963, pp. 337–340. Springer, Heidelberg (2008). https://doi.org/10.1007/978-3-540-78800-3_24
11. Selman, B., Kautz, H.A., Cohen, B.: Local search strategies for satisfiability testing. Cliques, Color. Satisf. **26**, 521–532 (1993)
12. Selman, B., Kautz, H.A., Cohen, B.: Noise strategies for improving local search. In: Proceedings of the 12th National Conference on Artificial Intelligence, vol. 94, pp. 337–343. MIT Press (1994)
13. Hoos, H.H., Stützle, T.: Stochastic Local Search: Foundations and Applications. Elsevier, Amsterdam (2004)
14. Gomes, C.P., Kautz, H., Sabharwal, A., Selman, B.: Satisfiability solvers. Foundations of Artificial Intelligence **3**, 89–134 (2008)
15. Kirschner, M., Mitchison, T.: Beyond self-assembly: from microtubules to morphogenesis. Cell **45**, 329–342 (1986)
16. Holy, T.E., Leibler, S.: Dynamic instability of microtubules as an efficient way to search in space. Proc. Nat. Acad. Sci. **91**, 5682–5685 (1994)
17. Gerhart, J., Kirschner, M.: Cells, Embryos, and Evolution: Toward a Cellular and Developmental Understanding of Phenotypic Variation and Evolutionary Adaptability. Blackwell Science, Malden (1997)
18. Kirschner, M., Gerhart, J.: Evolvability. Proc. Nat. Acad. Sci. **95**, 8420–8427 (1998)
19. Cauwenberghs, G.: A fast stochastic error-descent algorithm for supervised learning and optimization. In: Advances in Neural Information Processing Systems, pp. 244–251 (1993)
20. Sebastian Seung, H.: Learning in spiking neural networks by reinforcement of stochastic synaptic transmission. Neuron **40**, 1063–1073 (2003)
21. Hinton, G.E., Nowlan, S.J.: How learning can guide evolution. Complex Syst. **1**, 495–502 (1987)
22. Cook, M., Soloveichik, D., Winfree, E., Bruck, J.: Programmability of chemical reaction networks. In: Condon, A., Harel, D., Kok, J., Salomaa, A., Winfree, E. (eds.) Algorithmic Bioprocesses. Natural Computing Series, pp. 543–584. Springer, Heidelberg (2009)
23. Soloveichik, D., Seelig, G., Winfree, E.: DNA as a universal substrate for chemical kinetics. Proc. Nat. Acad. Sci. **107**, 5393–5398 (2010)
24. Chen, Y.-J.: Programmable chemical controllers made from DNA. Nat. Nanotechnol. **8**, 755 (2013)
25. Srinivas, N., Parkin, J., Seelig, G., Winfree, E., Soloveichik, D.: Enzyme-free nucleic acid dynamical systems. Science **358**, eaal2052 (2017)
26. Gillespie, D.T.: Stochastic simulation of chemical kinetics. Ann. Rev. Phys. Chem. **58**, 35–55 (2007)
27. Anderson, D.F., Cappelletti, D., Koyama, M., Kurtz, T.G.: Non-explosivity of stochastically modeled reaction networks that are complex balanced. Bull. Math. Biol. **80**, 2561–2579 (2018)

28. Kurtz, T.G.: The relationship between stochastic and deterministic models for chemical reactions. J. Chem. Phys. **57**, 2976–2978 (1972)
29. Karp, R.M., Miller, R.E.: Parallel program schemata. J. Comput. Syst. Sci. **3**, 147–195 (1969)
30. Johnson, R.F., Dong, Q., Winfree, E.: Verifying chemical reaction network implementations: a bisimulation approach. Theor. Comput. Sci. **765**, 3–46 (2019)
31. Doty, D., Zhu, S.: Computational complexity of atomic chemical reaction networks. Natural Comput. **17**, 677–691 (2018)
32. Magnasco, M.O.: Chemical kinetics is turing universal. Phys. Rev. Lett. **78**, 1190–1193 (1997)
33. Liekens, A.M.L., Fernando, C.T.: Turing complete catalytic particle computers. In: Almeida e Costa, F., Rocha, L.M., Costa, E., Harvey, I., Coutinho, A. (eds.) ECAL 2007. LNCS (LNAI), vol. 4648, pp. 1202–1211. Springer, Heidelberg (2007). https://doi.org/10.1007/978-3-540-74913-4_120
34. Soloveichik, D., Cook, M., Winfree, E., Bruck, J.: Computation with finite stochastic chemical reaction networks. Nat. Comput. **7**, 615–633 (2008)
35. Chen, H.-L., Doty, D., Soloveichik, D.: Deterministic function computation with chemical reaction networks. Natural Comput. **13**, 517–534 (2014)
36. Cummings, R., Doty, D., Soloveichik, D.: Probability 1 computation with chemical reaction networks. Natural Comput. **15**, 245–261 (2016)
37. Napp, N.E., Adams, R.P.: Message passing inference with chemical reaction networks. In: Advances in Neural Information Processing Systems, pp. 2247–2255 (2013)
38. Gopalkrishnan, M.: A scheme for molecular computation of maximum likelihood estimators for log-linear models. In: Rondelez, Y., Woods, D. (eds.) DNA 2016. LNCS, vol. 9818, pp. 3–18. Springer, Cham (2016). https://doi.org/10.1007/978-3-319-43994-5_1
39. Fett, B., Bruck, J., Riedel, M.D.: Synthesizing stochasticity in biochemical systems. In: Proceedings of the 44th Annual Design Automation Conference, pp. 640–645. ACM (2007)
40. Poole, W., et al.: Chemical boltzmann machines. In: Brijder, R., Qian, L. (eds.) DNA 2017. LNCS, vol. 10467, pp. 210–231. Springer, Cham (2017). https://doi.org/10.1007/978-3-319-66799-7_14
41. Cardelli, L., Kwiatkowska, M., Laurenti, L.: Programming discrete distributions with chemical reaction networks. Nat. Comput. **17**, 131–145 (2018)
42. Anderson, D.F., Craciun, G., Kurtz, T.G.: Product-form stationary distributions for deficiency zero chemical reaction networks. Bull. Math. Biol. **72**, 1947–1970 (2010)
43. Cappelletti, D., Ortiz-Munoz, A., Anderson, D., Winfree, E.: Stochastic chemical reaction networks for robustly approximating arbitrary probability distributions. arXiv preprint arXiv:1810.02854 (2018)
44. Braich, R.S., Chelyapov, N., Johnson, C., Rothemund, P.W.K., Adleman, L.: Solution of a 20-variable 3-SAT problem on a DNA computer. Science **296**, 499–502 (2002)
45. Kirkpatrick, S., Selman, B.: Critical behavior in the satisfiability of random Boolean expressions. Science **264**, 1297–1301 (1994)
46. Selman, B., Kirkpatrick, S.: Critical behavior in the computational cost of satisfiability testing. Artif. Intell. **81**, 273–295 (1996)
47. Strzebonski, A.: Mathematica SatisfiabilityQ uses MiniSAT 1.14. StackExchange and personal communication (2016, 2019). https://mathematica.stackexchange.com/questions/103726/why-is-satisfiabilitycount-faster-than-satisfiableq

48. Soloveichik, D.: CRNSimulatorSSA Mathematica Package. Personal Web Site (2016). http://users.ece.utexas.edu/~soloveichik/crnsimulator.html
49. Gibson, M.A., Bruck, J.: Efficient exact stochastic simulation of chemical systems with many species and many channels. J. Phys. Chem. A **104**, 1876–1889 (2000)
50. Mauch, S., Stalzer, M.: Efficient formulations for exact stochastic simulation of chemical systems. IEEE/ACM Trans. Comput. Biol. Bioinf. **8**, 27–35 (2011)
51. Thanh, V.H., Zunino, R.: Adaptive tree-based search for stochastic simulation algorithm. Int. J. Comput. Biol. Drug Des. **7**, 341–357 (2014)
52. Condon, A., Hu, A.J., Maňuch, J., Thachuk, C.: Less haste, less waste: on recycling and its limits in strand displacement systems. Interface Focus **2**, 512–521 (2012)
53. Thachuk, C., Condon, A.: Space and energy efficient computation with DNA strand displacement systems. In: Stefanovic, D., Turberfield, A. (eds.) DNA 2012. LNCS, vol. 7433, pp. 135–149. Springer, Heidelberg (2012). https://doi.org/10.1007/978-3-642-32208-2_11
54. Condon, A., Thachuk, C.: Towards space-and energy-efficient computations. In: Kempes, C., Grochow, J., Stadler, P., Wolpert, D. (eds.) The Energetics of Computing in Life and Machines, chapter 9, pp. 209–232. The Sante Fe Institute Press, Sante Fe (2019)
55. Matthew M Cook. Networks of Relations. PhD thesis, California Institute of Technology, 2005
56. Hopfield, J.J.: Neural networks and physical systems with emergent collective computational abilities. Proc. Nat. Acad. Sci. **79**, 2554–2558 (1982)
57. Kitano, H.: Towards a theory of biological robustness. Mol. Syst. Biol. **3**, 137 (2007)
58. Reich, E.S.: Mathematician claims breakthrough in Sudoku puzzle. Nature (2012). https://doi.org/10.1038/nature.2012.9751
59. Karr, J.R., Sanghvi, J.C., Macklin, D.N., Arora, A., Covert, M.W.: WholeCellKB model organism databases for comprehensive whole-cell models. Nucleic Acids Res. **41**, D787–D792 (2012)
60. Winfree, E.: Mathematica Notebooks for CRN SAT Solvers. Personal Web Site (2019). http://www.dna.caltech.edu/SupplementaryMaterial/CRNSAT/

Implementing Arbitrary CRNs Using Strand Displacing Polymerase

Shalin Shah[1]([✉]), Tianqi Song[2], Xin Song[1,3], Ming Yang[2], and John Reif[1,2]

[1] Department of Electrical and Computer Engineering, Duke University,
Durham, NC 27701, USA
shalin.shah@duke.edu
[2] Department of Computer Science, Duke University, Durham, NC 27701, USA
[3] Department of Biomedical Engineering, Duke University, Durham, NC 27701, USA
reif@cs.duke.edu

Abstract. The regulation of cellular and molecular processes typically involves complex biochemical networks. Synthetic nucleic acid reaction networks (both enzyme-based and enzyme-free) can be systematically designed to approximate sophisticated biochemical processes. However, most of the prior experimental protocols for reaction networks relied on either strand-displacement hybridization or restriction and exonuclease enzymatic reactions. These resulting synthetic systems usually suffer from either slow rates or leaky reactions. In this work, we propose an alternative architecture to implement arbitrary reaction networks, that is based entirely on strand-displacing polymerase reactions with non-overlapping I/O sequences. We first design a simple protocol that approximates arbitrary unimolecular and bimolecular reactions using polymerase strand displacement reactions. Then we use these fundamental reaction systems as modules to show three large-scale applications of our architecture, including an autocatalytic amplifier, a molecular-scale consensus protocol, and a dynamic oscillatory system.

Keywords: DNA polymerase · Strand-displacement ·
Chemical reaction networks · Consensus protocol · Chemical kinetics ·
Oscillatory protocols

1 Introduction

Ultra-high specificity and sensitivity of hybridization reactions by deoxyribonucleic acid (DNA) leads to a wide variety of useful applications [7,10,12,13,15, 16,18,19,21,26–28,30,31,37]. Particularly interesting use of DNA is to implement arbitrary chemical reaction networks [6,32,35]. Chemical Reaction Networks (CRNs) have been used to understand complex biochemical processes since they can represent biochemical systems in a simple-yet-formal way. CRNs are shown to be Turing universal [4,5,29] and therefore, in theory, can serve as a programming language for biochemical systems. A CRN consists of a set of reactant species ($R = \{R_1, R_2, ..., R_N\}$) with an initial stoichiometry. These reactants react via coupled chemical reactions to produce a set of product species ($P = \{P_1, P_2, ..., P_M\}$). Several dynamic phenomena have been experimentally described

C. Thachuk and Y. Liu (Eds.): DNA 25, LNCS 11648, pp. 21–36, 2019.
https://doi.org/10.1007/978-3-030-26807-7_2

and approximated using DNA as a substrate. For example, the molecular consensus protocol by Chen *et al.* implemented a set of CRNs using DNA to find the reactant species with the highest population [2,6]. Another interesting application includes implementation of the dynamic rock-paper-scissor oscillator by Srinivas *et al.* using DNA hybridization and strand displacement properties [35]. Other experimental demonstrations of dynamic system use either RNA polymerase and ribonuclease enzymes [17], or polymerase and nicking enzyme [11,20].

1.1 Motivation for Our Work

Since the early demonstrations of enzyme-free logic computing [25], the field of DNA nanoscience has mainly focused on using only DNA hybridization and toehold-mediated strand displacement as a fundamental unit for computing applications [3,6,7,15,23,40]. Although it is rational to use enzyme-free architectures because of their biological simplicity, the implementations of complex biochemical processes using enzyme-free toehold-mediated strand displacement reactions typically involve intricate sequence designs and additional gate complexes to mitigate problems such as reaction leak [35]. Several solutions have been proposed such as using clamps [39], shadow strands [33] and base-pair mismatches [14] to mitigate this issue. A general and practical way to mitigate such leaky reactions is using low concentrations. However, low concentration also reduces the overall rate of reaction or the speed of computation.

One way to address this challenge is to localize all the required computing DNA strands [1,3,36]. In a localization reaction, a breadboard-like structure such as a DNA nanotrack, DNA origami, and nanoparticles ensures the vicinity of the computing DNA strands. As all the DNA strands required for the reaction are placed in the close vicinity, the localized design no longer suffers from the slow kinetics of strand diffusion and unwanted DNA interactions. Although localization reaction reduced the reaction time and leaks, the overall design process becomes more complex as it requires design and synthesis of a nanostructure-like substrate [1].

Several other implementations of the biochemical reaction networks with much more sophisticated biological components have also been proposed [6,11,17,35]. For example, arbitrary biochemical systems can be implemented using the polymerase, nicking, and exonuclease (PEN) toolbox. However, by using the nicking enzyme, such protocols restrict the overall achievable reaction rates. This is mainly due to the slower activity of the nicking enzyme as compared to the polymerase enzyme [11,20].

1.2 Our Contribution

In this work, inspired by the success of enzyme-free DNA dynamic systems, we introduce a polymerase-based strand displacement architecture to implement arbitrary CRNs (refer to Fig. 1). We develop detailed designs of the template CRNs, namely, unimolecular and bimolecular reactions, using polymerase-based strand displacement. By doing so, we develop a generalized framework that can

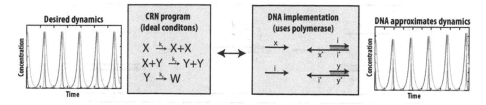

Fig. 1. The basic workflow of DNA systems implementing arbitrary reaction networks. The desired dynamic behavior can be represented using a set of CRNs. A DNA implementation of these CRNs can be tuned to approximately behave like the CRNs representing the biochemical process. The observed dynamic behavior of DNA systems mimics the original desired dynamics.

implement any arbitrary CRN since these reactions form the basis. There are several potential benefits of our CRN design framework. (a) It is simple and less leaky as compared to toehold-mediated strand displacement since no overlapping sequence between input and output signals exists, and the polymerase enzyme is sensitive to 3' overhangs. (b) It is faster than prior architectures which use polymerase, nicking and exonuclease together since our designs are polymerase only. (c) It is modular as we use template CRNs for large-scale applications such as an autocatalytic amplifier and molecular consensus network. To demonstrate arbitrary CRNs, we approximate a dynamic oscillator with polymerase-based DNA design as this requires a very careful calibration of the reaction rates. All our designs are demonstrated *in silico* using a system of coupled ordinary differential equations (ODEs) as it is well-known that the time evolution of a CRN can be specified by a system of ODEs [1,6,32,40].

1.3   Paper Organization

The rest of the paper is organized as follows: First, we introduce polymerase-based strand displacement and compare it with the enzyme-free toehold-mediated strands displacement. In particular, our protocols only use hybridization and polymerase-based strand displacement reactions, and do not use enzyme-free toehold-mediated strands displacement reactions. Second, we introduce two fundamental CRN systems, namely, unimolecular reaction and bimolecular reaction, as they form the basis for any complex biochemical process. Here, we argue how our DNA design can implement these fundamental reactions. Third, using these simple CRNs as a template, we demonstrate three different applications: (a) an autocatalytic amplifier, (b) a molecular-scale consensus system, and (c) a dynamic rock-paper-scissor oscillatory system. Finally, we conclude our paper with a discussion on experimental demonstration and other future work.

2   Strand Displacement with DNA Polymerase

The basic principles behind toehold-mediated strand displacement as well as polymerase-based strand displacement are shown in Fig. 2 for comparison.

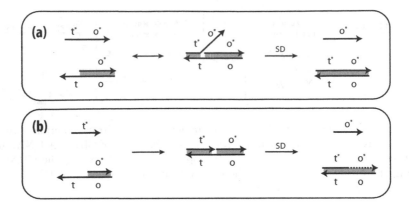

Fig. 2. The basic mechanism of toehold-mediated strand displacement and polymerase-based strand displacement. (a) In toehold-mediated strand displacement process, an incoming signal strand has a short toehold domain that reversibly binds with the complex until the branch migration step happens and the output signal is completely displaced from the complex. The toehold domain is usually kept short (approx. under 10 nt) while the branch is kept longer (approx. 15 nt or more). (b) In polymerase-based strand displacement process, an incoming signal strand permanently binds with the complementary domain on the complex at which point the polymerase enzyme kicks in to displace output strand. All the domains such as t and o can be designed long (approx. 30 nt) as the polymerase enzyme works at higher temperatures. SD refers to strand displacement.

In toehold-mediated strand displacement, an input signal contains a short domain called the toehold which can hybridize with the exposed region of a double-stranded DNA complex. Once the input signal binds with the complex, it undergoes a back and forth tug-of-war (referred to as the branch migration step) until the incumbent output strand attached to the complex undergoes complete dehybridization. The initial attachment of the input signal to the complex is a reversible process as the toehold domains are usually kept short. However, they can be made longer to increase the overall rate of the process. In polymerase-based strand displacement, the input signal is designed such that it comes in and binds with the complex permanently, as shown in Fig. 2b. Once bound, it can act as a primer site for the polymerase enzyme. The enzyme attaches to the DNA complex to begin its polymerase activity, from the 5' to 3' direction, along with the displacement of the incumbent strand. The dotted strand in Fig. 2b indicates the priming activity of the polymerase enzyme. The benefit of using polymerase-based implementation is the nonoverlapping DNA sequence between the input and output signals, which is known to produce reaction leaks [39]. Several strand displacement polymerase enzymes are commercially available such as Bst and Phi providing a wide range of available experimental conditions.

3   Implementation of Arbitrary Reactions

Unimolecular and bimolecular reactions are the most widely studied class of chemical reactions as more complex reactions can be easily broken down into these fundamental reactions [6,8,32,35]. This makes them the basis for implementing arbitrary complex reaction networks. Therefore, in this work, we first focus on implementing arbitrary unimolecular and bimolecular reactions using DNA hybridization and polymerase-based strand displacement as the fundamental building blocks.

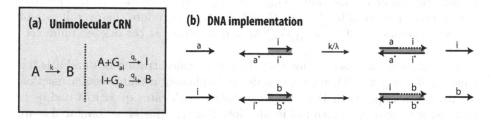

Fig. 3. Arbitrary unimolecular reaction implementation using polymerase-based DNA strand displacement. (a) An arbitrary transducer CRN network that absorbs A and emits B in a two-step process. Input A combines with gate G_{ai} to release intermediate I which then combines with gate G_{ib} to release output B. Both the reactions use strand displacing polymerase enzyme to facilitate the reaction. (b) A domain-level implementation of the DNA-based CRN which shows the hybridization, priming, and polymerization process. The rate of the second step is not shown as it is much faster and therefore occurs instantaneously w.r.t to the first step.

3.1   Arbitrary Unimolecular Reactions

The simplest CRN system is a signal transducer with applications in the field of cell biology, molecular biology, and electronics. In this system, an input signal A is absorbed to produce a different type of signal B as its output, written in a simplified form as A $\xrightarrow{k}$ B. An enzyme-free implementation of such arbitrary unimolecular CRN was first proposed by Soloveichik *et al.* [32]. Here, we report a simple two-step design that uses the polymerase-based strand displacement and DNA hybridization as its fundamental units. We implement the unimolecular CRN using a two-step process to avoid complications with sequence design associated with single-step implementations, especially if the input and output share the same domain.

An arbitrary unimolecular CRN A $\xrightarrow{k}$ B can be implemented using two gate complexes G_{ai} and G_{ib}, as shown in Fig. 3. The input signal A can act as a primer upon hybridization with complex G_{ai} for the polymerase reaction, leading to the displacement of incumbent intermediate strand I. This intermediate strand I can hybridize with the output complex G_{ib} to undergo another polymerase-based strand displacement to release the output B. The dotted strands in the figure indicate the polymerase activity of the DNA polymerase enzyme.

For this two-step DNA implementation to approximate a first order reaction, the overall rate of reaction should be k[A]. This can be achieved by careful calibration of reaction conditions. Let us assume that the gate concentrations are much higher w.r.t input strand *i.e* $[G_{ai}] >> [A_0]$ and $[G_{ib}] >> [A_0]$ making their effective concentrations constant throughout the reaction duration. By doing so both the reactions become pseudo-first order reactions given by $A \xrightarrow{q_1[G_{ai}]} I$ and $I \xrightarrow{q_2[G_{ib}]} B$. Then to make the overall reaction approach first-order kinetics, we simply need to make sure intermediate complex I is in quasi-steady-state equilibrium i.e. it gets consumed immediately after production. We can easily make second step faster by assuming $[G_{ib}] >> [G_{ai}]$. At the steady state, the concentration of I is given by $[I] = q_1[G_{ai}][A]/q_2[G_{ib}]$, therefore the overall rate of reaction becomes $q_2[I][G_{ib}] \approx q_1[G_{ai}][A] \approx q[A]$ acting as the desired unimolecular reaction.

For simplicity, we also assume that the excess concentration is λ and it is the same for all the gates. Therefore, in order to implement a unimolecular reaction of rate k, we simply need DNA implementation with rates $q_1 = k/\lambda$ and q_2 is very fast w.r.t to q_1 as shown in Fig. 3b. Note that q_2 is faster as compared to q_1 hence irrelevant and not shown. For more details on steady state approximation and other theoretical proofs, refer to the prior work by Soloveichik *et al.* [32].

The rate of reactions in the DNA implementation can be tuned either by changing the gate concentration or by tuning the rate of strand displacement. However, throughout this work, we assume that the gate concentrations are constant and the programmable behavior comes from the tunable rate of strand displacement. This can be achieved by adjusting the length of the primer and its GC content [11,34,35]. Note that the output complex can be easily changed to incubate one, two or more strands of either same or different types. Therefore, it can be easily tuned to implement a catalytic($A \xrightarrow{k} A + B$) and autocatalytic ($A \xrightarrow{k} A + A$) reaction, as we will see in the applications section (see Fig. 6).

3.2   Arbitrary Bimolecular Reactions

The simple two-step process used for the unimolecular reaction can be easily extended to implement a bimolecular reaction. In bimolecular reaction, two set of inputs A and B combine to produce an output C such that the rate of output formation depends on the inputs as k[A][B]. A trivial implementation of bimolecular reaction would be to slightly modify unimolecular to design to replace intermediate domain i by input domain b. Then, a two-step reaction $A + G_b \xrightarrow{q_1} I$ and $I + G_{io} \xrightarrow{q_2} O$ similar to previous design can act as a bimolecular reaction. This trivial design has an issue that it expects second input b to be present in the gated form. This can create a problem, especially, if there are multiple reactions that use b as the input signal. For example, if our system of CRNs contains $A + B \xrightarrow{k_1} C$ and $B \xrightarrow{k_2} D$ then the input signal B is required in a single-stranded form in the unimolecular reaction and in the double-stranded form in the bimolecular reaction. This problem was originally also identified by Solove-

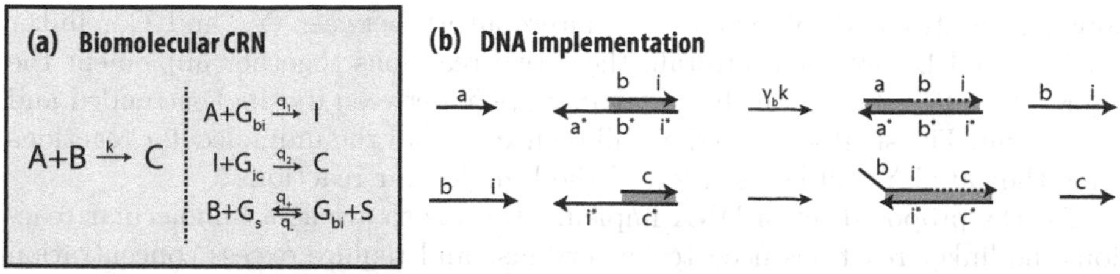

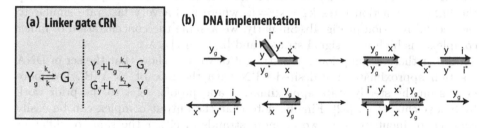

(a) Biomolecular CRN

$$A+B \xrightarrow{k} C$$

$$A+G_{bi} \xrightarrow{q_i} I$$
$$I+G_{ic} \xrightarrow{q_i} C$$
$$B+G_s \xrightleftharpoons[q_i]{q_i} G_{bi}+S$$

(b) DNA implementation

Fig. 4. Arbitrary bimolecular reaction implementation using polymerase-based DNA strand displacement. (a) An arbitrary non-catalytic CRN network that absorbs A and B in a two-step process to emit C. Input A combines with gate G_{bi} to releases intermediate I which then combines with gate G_{ic} to release output C. A linker reaction reversibly converts input B to gated form G_{bi}. DNA implementation details of this gate are shown in Fig. 5(b) A domain-level implementation of the DNA-based CRN. Unlabeled rate constant is fast and therefore not shown.

(a) Linker gate CRN

$$Y_g \xrightleftharpoons[k]{k_+} G_y$$

$$Y_g+L_g \xrightarrow{k_1} G_y$$
$$G_y+L_y \xrightarrow{k_2} Y_g$$

(b) DNA implementation

Fig. 5. Linker gates to interconvert a signal strand. (a) A CRN network that reversibly converts input signal Y between the gated-form and the single-stranded form. The single-stranded signal Y_g is used in all the unimolecular reactions of Y while the gate complex G_y will be used as input in all the bimolecular reactions. (b) A domain-level implementation of the DNA-based CRN. All the reactions are relatively fast and therefore rate constants are not shown.

ichik *et al.* [32] when they designed a DNA-only CRN system and they proposed the use of the well-known toehold exchange reaction to reversibly interconvert input signal B between its two forms [8, 24].

In this work, we design a simple set of strand displacement reactions to implement interconversion of input signal B between its single-stranded and gated form. Therefore, the final set of CRNs that implements a bimolecular reaction will consist of the third linker step in addition to the two-step process, as shown in Figs. 4a and 5. The input signal A can bind with gate G_{bi} and undergo polymerase-based strand displacement to release the intermediate I. This intermediate strand I binds with the output complex G_{ic} to release output C. The additional linker step includes reversible reaction whose full implementation is shown in Fig. 5.

Abstractly the linker reaction converts single-stranded input signal Y_g with help of linker complexes to gate G_y in a reversible fashion. The polymerase implementation of this CRN is a two-step process where each step uses only polymerase strand displacement. To interconvert between Y_g and G_y, linker gates L_g and L_y are used. Overall, these two reactions together implement the reversible buffered reaction that interconverts Y between its single-stranded and gated form. The single-stranded Y will be used for all the unimolecular reactions while the gated Y will be used for all the bimolecular reactions.

For the proposed set of DNA implementations to act as a bimolecular reaction, the linker reactions need to be very fast and require excess concentration of all the gate complexes. This includes the two linker gates L_g and L_y, and the gate complex G_{ic}. The actual available concentration of the input signal B (or Y) used in both the unimolecular and bimolecular reactions will be half assuming the linker reaction reaches a quasi-steady-state on a fast time scale w.r.t other reactions. This means the total concentration of B is $[B_T] = [B] + [G_B]$. If e(B) is the equilibrium concentration of B, then under the pseudo-equilibrium assumption, $e(B) = [B]/[B_T]$. Therefore, we multiply the rate of reaction by a scaling factor of $\gamma = 1/e(B)$ to account for this activity. For simplicity, we keep the buffered reaction rates $k_+ = k_- = k$, where k is a very fast rate similar to the second reaction in Fig. 4b. Similarly, we assume the concentration of linker complexes is high w.r.t signal strands and $[L_g] = [L_y] = \lambda$.

Under these assumptions, it is straightforward to show that the set of DNA reaction approximates the desired CRN with the rate of $\gamma k[A][B]$. For more details on the steady state approximation and proofs, refer to the prior work by Soloveichik *et al.* [32]. Finally, note that the output complex can be easily changed to incubate one, two or more strands of either the same or different types. Therefore, it can be easily tuned to implement catalytic($A + B \xrightarrow{k} B + C$) and autocatalytic ($A + B \xrightarrow{k} C + B + B$) reactions, as we will see in the applications section.

4   Scaling Reaction Systems for Practicality

All the DNA implementations of the CRN protocols in this work are scaled by a time factor α and a concentration factor β as the DNA implementations only work in a certain range of time and concentration. A standard DNA hybridization experiment operates in the concentration range of 1 nM to 1 mM and in the time range of a few minutes to several hours [1, 6, 11, 23, 25, 34, 36, 39]. Therefore, a set of coupled ODEs implementing a CRN system with rate $k = 1$ can be multiplied by a time factor $\alpha = 10^{-3}$ and a concentration factor $\beta = 10^{-8}$. This will bring our DNA implementation of the CRN systems in a realistic range without altering the behavior of the original system [32]. Unless otherwise stated all the applications will use these values of k, α and β for simplicity. All the scaling factors and other rate values are adopted from Soloveichik *et al.* [32].

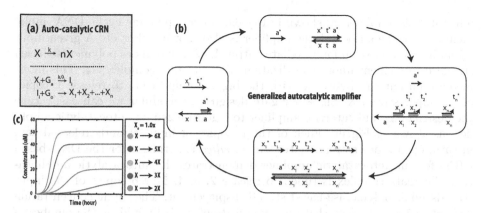

Fig. 6. A generalized autocatalytic amplifier using polymerase-based strand displacement. (a) The set of CRNs for an autocatalytic amplifier and CRNs for the DNA implementation. (b) The flow of an autocatalytic reaction. The input X opens the first gate to release intermediates which open the second gate to release multiple copies of X. All the subscripts just indicate the number of times a domain is repeated. Here the number of domains in the complexes can be tuned to adjust the amplification factor n. (c) An *in silico* demonstration of the 2x, 3x, 4x, 5x and 6x amplifier. Excess concentration of the supporting gates is $\lambda = 10^{-5}$ M while the unlabeled fast reaction rates were 10^5.

5 Applications

The simple CRNs introduced in the previous sections can act as a basis for large-scale applications. In this section, we use these simple reactions as a template and demonstrate several applications as a use case of our polymerase-based architecture.

5.1 Generalized Autocatalytic Amplifier

Autocatalytic amplification is one of the simplest and most widely used tasks in the field of molecular biology and DNA nanoscience. Such an endpoint system does not require very intricate control of reaction rates and input concentration making them the simplest to implement. While most amplification techniques require either programming multiple thermal cycles to achieve an exponential amplification or a nicking enzyme, our design of autocatalytic amplifier works at the same temperature and requires only polymerase enzyme [22,38]. The unimolecular reaction designed in the previous section can be easily tuned to design a fast autocatalytic amplifier by modification of the gate strand. Such autocatalytic amplifier can also serve as a signal restoration module in complex DNA-based digital logic circuits [7] where the output signal decays in each circuit layer.

The gate complex G_{ib} used in the unimolecular CRN design, shown in Fig. 3, can be tuned to release multiple copies of B (or A) and implement a generalized autocatalytic amplifier, as shown in Fig. 6a. A corresponding cyclic DNA implementation of the autocatalytic amplifier is also shown in Fig. 6b. A given input signal X_i can release an intermediate output I_i as it undergoes polymerase-based strand displacement upon hybridization with the gate complex G_a. This intermediate output can hybridize with the bigger complex G_o to release k copies of the input strand X. Since our gate designs are modular we can easily tune the design of 2x autocatalytic amplifier to a generalized kx autocatalytic amplifier, as shown in Fig. 6b. Each of the k copies of input strands released can, in turn, open k downstream gates. An *in silico* experiment for the DNA-based CRNs for our autocatalytic amplifier demonstrates the autocatalytic amplification. The gate complex is tuned to release 2, 3, 4, 5 or 6 copies of the input strands upon polymerase-based strand displacement. The input concentration was set to 100 nM. Note that our amplified output is limited by the number of available gate complexes as seen in Fig. 6c.

5.2 Molecular-Scale Consensus Network

We next try a more complex endpoint application to demonstrate the generality of our polymerase-based architecture. An interesting problem in the field of distributed computing and contemporary economics is for the network to reach a consensus. Consensus algorithms can help to decide, for example, the democratic leader by increasing the population of the party in the majority while reducing the population of the party in the minority. Such algorithms are generally required in distributed systems to make them fault tolerant. The first molecular-scale consensus network was demonstrated by Chen *et al.* using toehold-mediated strand displacement [6]. In this work, we tune our template reactions to design a DNA polymerase-based consensus network.

A CRN-based consensus protocol autonomously identifies the reactant with the highest initial concentration and increases its concentration to be much higher. The other reactant with lower initial concentration is decreased close to zero, as shown in Fig. 7a. The consensus protocol consists of three autocatalytic reactions $A + B \xrightarrow{k_1} 2Y$, $B + Y \xrightarrow{k_2} 2B$ and $A + Y \xrightarrow{k_3} 2A$. A set of abstract CRNs along with DNA-based CRNs that can implement this network is shown in Fig. 7b. Note that we have omitted the domain-level DNA representation of this CRN for brevity, however it is similar to the template reactions shown in Figs. 3 and 4. The gate Y was kept at 30 nM concentration while the relative concentration of input strands A and B were changed 1 to 30 nM (1x = 30 nM) to test our design of the consensus protocol. The rate constant of each bimolecular reaction is multiplied by $\gamma = 2$ to account for the linker effect. All the rate constants not shown are faster as compared to the main reaction and therefore not shown. The supporting gates with very high concentration were $\lambda = 10^{-5}$ M while the unlabeled fast reaction rates were 10^3. The relative concentration of each input strand was modified from 0.1x to 0.9x as shown in Fig. 7c. Clearly, as seen from the *in silico* experiments, the consensus CRN algorithm can absorb the

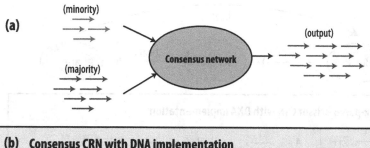

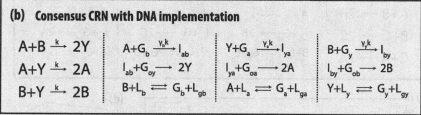

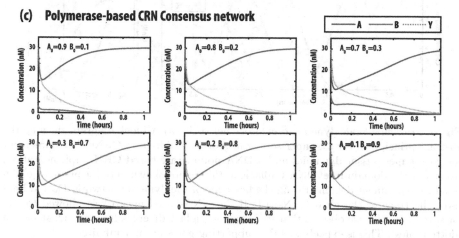

Fig. 7. Molecular-scale polymerase-based consensus network. (a) A block diagram showing the abstract principle of a consensus network. When multiple species are present, the algorithm automatically finds the majority by reducing the minority species close to zero. (b) Abstract ideal CRNs that implement the consensus algorithm and a DNA-polymerase based CRN implementation. Note that the domain-level DNA implementation is not shown here for brevity. (c) *In silico* demonstration of the consensus network at different relative concentrations of inputs A and B. The gate Y was kept at 30 nM concentration while the inputs A and B were varied from 3 nM to 27 nM (1x = 30 nM) to change the relative concentration. Excess concentration of the supporting gates is $\lambda = 10^{-5}$ M while the unlabeled fast reaction rates were 10^3.

species whose population is a minority bringing its concentration down nearly to zero. The species in majority are consumed initially but eventually stops decreasing indicating a democratic decision is made by the molecular network.

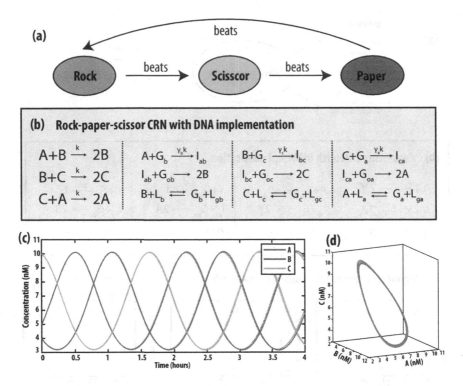

Fig. 8. Molecular-scale dynamic oscillatory network. (a) A block diagram showing the abstract principle of a rock-paper-scissor network. (b) Abstract CRNs that implement the rock-paper-scissor dynamics and a DNA polymerase-based CRN implementation. Note that the domain-level DNA implementation is not shown here for brevity. (c) An *in silico* experiment demonstrating the behavior of the dynamic network. Dotted lines show the trace of the ideal CRNs. (d) If the DNA-based network is allowed to run for several epochs, we observe that its behavior slightly diverges from the ideal system (dotted lines). This is expected as the supporting gates are not infinite.

5.3   Molecular-Scale Dynamic Oscillator

The final application we demonstrate here is the well-known dynamic rock-paper-scissor oscillator. Such oscillatory protocols have much more stringent requirements on the initial conditions and the rate constants than endpoint systems such as an autocatalytic amplifier or the consensus protocol [6,32,35]. A successful DNA implementation of such systems demonstrates a strong use case of the proposed polymerase architecture. Prior enzyme-free architecture by Srinivas *et al.* reported the first DNA implementation of this oscillator [35]. Other enzyme-based oscillator systems have also been reported [11,17]. Here we explore an alternate avenue by designing a DNA polymerase only implementation of this dynamic system.

The basic principle of such a system is to mimic the zero-sum hand game rock-paper-scissors, where there is no single winner. Rock can beat scissors by breaking it, scissors can beat paper by cutting it and paper can beat rock by covering it. This forms a cyclic system as shown in Fig. 8a whose CRN representation is shown in Fig. 8b. Three bimolecular autocatalytic reactions $(A + B \xrightarrow{k_1} 2B, B + C \xrightarrow{k_2} 2C, C + A \xrightarrow{k_3} 2A)$ are required to implement such a dynamic system. The first CRN produces B by consuming A. By assuming A and B as scissor and rock respectively, such CRN implements rock beating scissor by producing more rock while consuming scissor. Similarly, other CRNs implement scissor beats paper and paper beats rock respectively. Using our template reactions, we can easily implement this CRN system. The CRN for DNA implementation of each bimolecular reaction is shown in Fig. 8b.

An *in silico* demonstration of the dynamic system implemented using our DNA polymerase-based framework is shown in Fig. 8c. Each component of the CRN exists in gated form and single-stranded form through a reversible linker reaction. Thus $\gamma = 2$ for each reaction as all the inputs are used once in unimolecular form and once in bimolecular form. Initial input concentrations were 5 nM, 4 nM and 10 nM. All the supporting gates were kept at an excess concentration of $\lambda = 10^{-3}$M and the unlabeled fast reaction rates were 10^3. Initially, the DNA implementation exactly mimics the ideal CRNs (shown as dotted line in Fig. 8c), however as time progresses, it starts diverging away (refer Fig. 8d) since the supporting gates get consumed and therefore effective reaction rates change.

6   Discussion

6.1   Experimental Demonstration and Considerations

The first step towards experimental demonstration includes a careful study of the polymerase-based strand displacement kinetics to develop a reliable biophysical model. This includes understanding the influence of the primer and branch length, enzyme concentration, temperature etc. on the overall rate of a signal transducer reaction. A careful understanding of polymerase-based reactions also involves the study of leaky reactions. Several control experiments can be conducted with the supporting gates to identify the initial leak and gradual leaks [35]. However, the longer domain lengths, simple gate design, nonoverlapping I/O sequence and sensitivity of the polymerase enzyme to 3' overhangs should potentially mitigate the observed reaction leaks. Lower leaks can also allow the use of higher input concentrations enabling a much faster rate of reaction. The computing speed of our architecture can be compared to prior architectures using two different parameters: (a) half-time to reaction completion [1], (b) estimation of the rate constant [9].

6.2   Replenishing Supporting Gates with Buffered Reaction

Our DNA architecture heavily relies on the supporting and linker gates for approximating ideal CRNs. However, for example, in contrast to the prior

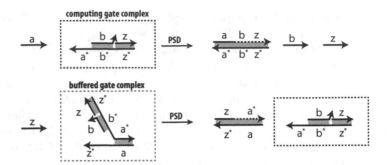

Fig. 9. Input signal A primes with the gate complex to produce output signal B and Z. The computing gate can be replenished by adding the buffered gate shown in the second reaction. All the supporting gates used in our architecture can be replenished by adding several such buffered complexes. This can autonomously elongate the longevity of the DNA systems. (Color figure online)

scheme [32], our linker gate design is wasteful. As the linker and other gates are consumed, the DNA system starts diverging away from the desired dynamics and eventually stops working. Although we can start with a higher initial concentration, it is not very desirable since several unwanted reactions occur at higher concentrations. Upon their consumption, either more gates should be added externally or a simple buffered gate scheme, as shown in Fig. 9, can be used. In such buffered systems [24], an additional gate complex (marked in the red box) will release the used gate (marked in the black box) upon its consumption. Such a system can easily elongate the longevity of our DNA systems as the supporting gates get replenished autonomously.

7   Conclusion

In this work, we introduced a DNA polymerase-based architecture to implement arbitrary chemical reaction networks. First, we demonstrated a DNA-based implementation of arbitrary unimolecular and bimolecular reactions as these form the basis for any complex system. We then used these reactions as a template to demonstrate several applications such as autocatalytic amplifier, consensus network, and dynamic rock-paper-scissor oscillator as strong use cases of our architecture. Our designs used only the polymerase enzyme and DNA hybridization making them simple to design and implement. The template reactions could be easily tuned to implement arbitrary biochemical reaction networks demonstrating the modularity of our framework.

Acknowledgements. This work was supported by National Science Foundation Grants CCF-1813805 and CCF-1617791. The authors thank Keerti Anand for useful theoretical discussions on ODEs and chemical kinetics. The most up-to-date simulation scripts are available online as a GitHub repository at https://bit.ly/2X3axAh.

References

1. Bui, H., Shah, S., Mokhtar, R., Song, T., Garg, S., Reif, J.: Localized DNA hybridization chain reactions on DNA origami. ACS Nano **12**(2), 1146–1155 (2018)
2. Cardelli, L., Csikász-Nagy, A.: The cell cycle switch computes approximate majority. Sci. Rep. **2**, 656 (2012)
3. Chandran, H., Gopalkrishnan, N., Phillips, A., Reif, J.: Localized hybridization circuits. In: Cardelli, L., Shih, W. (eds.) DNA 2011. LNCS, vol. 6937, pp. 64–83. Springer, Heidelberg (2011). https://doi.org/10.1007/978-3-642-23638-9_8
4. Chen, H.L., Doty, D., Soloveichik, D.: Deterministic function computation with chemical reaction networks. Nat. Comput. **13**(4), 517–534 (2014)
5. Chen, H.L., Doty, D., Soloveichik, D.: Rate-independent computation in continuous chemical reaction networks. In: Proceedings of the 5th Conference on Innovations in Theoretical Computer Science, pp. 313–326. ACM (2014)
6. Chen, Y.J., et al.: Programmable chemical controllers made from DNA. Nat. Nanotechnol. **8**(10), 755–762 (2013)
7. Cherry, K.M., Qian, L.: Scaling up molecular pattern recognition with DNA-based winner-take-all neural networks. Nature **559**(7714), 370 (2018)
8. Dalchau, N., et al.: Computing with biological switches and clocks. Nat. Comput. **17**(4), 761–779 (2018)
9. Eshra, A., Shah, S., Song, T., Reif, J.: Renewable DNA hairpin-based logic circuits. IEEE Trans. Nanotechnol. **18**, 252–259 (2019)
10. Fu, D., Shah, S., Song, T., Reif, J.: DNA-based analog computing. In: Braman, J.C. (ed.) Synthetic Biology. MMB, vol. 1772, pp. 411–417. Springer, New York (2018). https://doi.org/10.1007/978-1-4939-7795-6_23
11. Fujii, T., Rondelez, Y.: Predator-prey molecular ecosystems. ACS Nano **7**(1), 27–34 (2012)
12. Garg, S., Shah, S., Bui, H., Song, T., Mokhtar, R., Reif, J.: Renewable time-responsive DNA circuits. Small **14**(33), 1801470 (2018)
13. Han, D., et al.: Single-stranded DNA and RNA origami. Science **358**(6369), eaao2648 (2017)
14. Jiang, Y.S., Bhadra, S., Li, B., Ellington, A.D.: Mismatches improve the performance of strand-displacement nucleic acid circuits. Angew. Chem. **126**(7), 1876–1879 (2014)
15. Joesaar, A., et al.: DNA-based communication in populations of synthetic protocells. Nat. Nanotechnol. **14**, 369 (2019)
16. Johnson-Buck, A., Shih, W.M.: Single-molecule clocks controlled by serial chemical reactions. Nano Lett. **17**(12), 7940–7944 (2017)
17. Kim, J., Winfree, E.: Synthetic in vitro transcriptional oscillators. Mol. Syst. Biol. **7**(1), 465 (2011)
18. Kishi, J.Y., Schaus, T.E., Gopalkrishnan, N., Xuan, F., Yin, P.: Programmable autonomous synthesis of single-stranded DNA. Nat. Chem. **10**(2), 155 (2018)
19. Li, J., Johnson-Buck, A., Yang, Y.R., Shih, W.M., Yan, H., Walter, N.G.: Exploring the speed limit of toehold exchange with a cartwheeling DNA acrobat. Nat. Nanotechnol. **13**(8), 723 (2018)
20. Montagne, K., Plasson, R., Sakai, Y., Fujii, T., Rondelez, Y.: Programming an in vitro DNA oscillator using a molecular networking strategy. Mol. Syst. Biol. **7**(1), 466 (2011)
21. Newman, S., et al.: High density DNA data storage library via dehydration with digital microfluidic retrieval. Nat. Commun. **10**(1), 1706 (2019)

22. Notomi, T., et al.: Loop-mediated isothermal amplification of DNA. Nucleic Acids Res. **28**(12), e63–e63 (2000)
23. Qian, L., Winfree, E.: Scaling up digital circuit computation with DNA strand displacement cascades. Science **332**(6034), 1196–1201 (2011)
24. Scalise, D., Dutta, N., Schulman, R.: DNA strand buffers. J. Am. Chem. Soc. **140**(38), 12069–12076 (2018)
25. Seelig, G., Soloveichik, D., Zhang, D.Y., Winfree, E.: Enzyme-free nucleic acid logic circuits. Science **314**(5805), 1585–1588 (2006)
26. Seo, J., Kim, S., Park, H.H., Nam, J.M., et al.: Nano-bio-computing lipid nano-tablet. Sci. Adv. **5**(2), eaau2124 (2019)
27. Shah, S., Dubey, A., Reif, J.: Programming temporal DNA barcodes for single-molecule fingerprinting. Nano Lett. **19**, 2668–2673 (2019)
28. Shah, S., Dubey, A.K., Reif, J.: Improved optical multiplexing with temporal DNA barcodes. ACS Synth. Biol. **8**(5), 1100–1111 (2019)
29. Shah, S., Gupta, M.: DNA-based chemical compiler (2018). arXiv preprint arXiv:1808.04790
30. Shah, S., Limbachiya, D., Gupta, M.K.: DNACloud: a potential tool for storing big data on DNA (2013). arXiv preprint arXiv:1310.6992
31. Shah, S., Reif, J.: Temporal DNA barcodes: a time-based approach for single-molecule imaging. In: Doty, D., Dietz, H. (eds.) DNA 2018. LNCS, vol. 11145, pp. 71–86. Springer, Cham (2018). https://doi.org/10.1007/978-3-030-00030-1_5
32. Soloveichik, D., Seelig, G., Winfree, E.: DNA as a universal substrate for chemical kinetics. Proc. Nat. Acad. Sci. **107**(12), 5393–5398 (2010)
33. Song, T., et al.: Improving the performance of DNA strand displacement circuits by shadow cancellation. ACS Nano **12**(11), 11689–11697 (2018)
34. Srinivas, N., et al.: On the biophysics and kinetics of toehold-mediated DNA strand displacement. Nucleic Acids Res. **41**(22), 10641–10658 (2013)
35. Srinivas, N., Parkin, J., Seelig, G., Winfree, E., Soloveichik, D.: Enzyme-free nucleic acid dynamical systems. Science **358**(6369), eaal2052 (2017)
36. Teichmann, M., Kopperger, E., Simmel, F.C.: Robustness of localized DNA strand displacement cascades. ACS Nano **8**(8), 8487–8496 (2014)
37. Thubagere, A.J., et al.: A cargo-sorting DNA robot. Science **357**(6356) (2017)
38. Walker, G.T., Little, M.C., Nadeau, J.G., Shank, D.D.: Isothermal in vitro amplification of DNA by a restriction enzyme/DNA polymerase system. Proc. Nat. Acad. Sci. **89**(1), 392–396 (1992)
39. Wang, B., Thachuk, C., Ellington, A.D., Winfree, E., Soloveichik, D.: Effective design principles for leakless strand displacement systems. Proc. Nat. Acad. Sci. **115**(52), E12182–E12191 (2018)
40. Yordanov, B., Kim, J., Petersen, R.L., Shudy, A., Kulkarni, V.V., Phillips, A.: Computational design of nucleic acid feedback control circuits. ACS Synth. Biol. **3**(8), 600–616 (2014)

Real-Time Equivalence of Chemical Reaction Networks and Analog Computers

Xiang Huang[1], Titus H. Klinge[2], and James I. Lathrop[1]($\boxtimes$)

[1] Iowa State University, Ames, IA 50011, USA
{huangx,jil}@iastate.edu
[2] Drake University, Des Moines, IA 50311, USA
titus.klinge@drake.edu

Abstract. This paper investigates the class $\mathbb{R}_{RTCRN}$ of real numbers that are computable in real time by chemical reaction networks (Huang, Klinge, Lathrop, Li, Lutz, 2019), and its relationship to general purpose analog computers. Roughly, $\alpha \in \mathbb{R}_{RTCRN}$ if there is a chemical reaction network (CRN) with integral rate constants and a designated species X such that, when all species concentrations are initialized to zero, X converges to α exponentially quickly. In this paper, we define a similar class $\mathbb{R}_{RTGPAC}$ of real numbers that are computable in real time by general purpose analog computers, and show that $\mathbb{R}_{RTGPAC} = \mathbb{R}_{RTCRN}$ using a construction similar to that of the difference representation introduced by Fages, Le Guludec, Bournez, and Pouly. We prove this equivalence by showing that $\mathbb{R}_{RTCRN}$ is a subfield of $\mathbb{R}$ which solves a previously open problem. We also prove that a CRN with integer initial concentrations can be simulated by a CRN with all zero initial concentrations. Using these results, we give simple and natural constructions showing e and π are members of $\mathbb{R}_{RTCRN}$, which was not previously known.

1 Introduction

Computing real numbers in real time with a Turing machine was first introduced by Yamada in the 1960s [22] and was later proven to be equivalent to producing the first n bits of the fractional part of the number in $O(n)$-time [15]. It is easy to see that the rational numbers can be computed in real time, but others are not as trivial to compute. Some transcendental numbers are known to be real time computable, such as

$$\lambda = \sum_{n=1}^{\infty} 2^{-n!},$$

which is a Liouville number [18]. Surprisingly, the long standing conjecture by Hartmanis and Stearms that no irrational algebraic number can be computed in real time remains unsolved [18]. If the conjecture is true, it would imply that certain transcendental numbers are easier to compute by a Turing machine than algebraic numbers, which would be surprising indeed.

This research was supported in part by National Science Foundation Grant 1545028.

C. Thachuk and Y. Liu (Eds.): DNA 25, LNCS 11648, pp. 37–53, 2019.
https://doi.org/10.1007/978-3-030-26807-7_3

Recently, Huang, Klinge, Lathrop, Li, and Lutz defined the class $\mathbb{R}_{RTCRN}$ of real time computable real numbers by chemical reaction networks [19]. They defined a real number α to be in $\mathbb{R}_{RTCRN}$ if there is a CRN with a designated species X that is within 2^{-t} of $|\alpha|$ for all $t \geq 1$. Their definition also requires that all rate constants of the CRN be integral, all species concentrations be initialized to zero, and all species concentrations be bounded by some constant β. This definition is closely related to that of Bournez, Fraigniaud, and Koegler's large population protocols [3], but differs in key technical aspects. A discussion of these differences can be found in [19].

Common variants of the chemical reaction network model assume well-mixed solutions and mass-action kinetics. Small-volume environments such as *in vivo* reaction networks are often modeled stochastically with continuous-time Markov processes, while larger volumes are modeled deterministically with systems of ordinary differential equations[1]. In this paper, we focus on the deterministic mass-action chemical reaction network which is modeled with autonomous polynomial differential equations. Roughly, a chemical reaction network (CRN) is a set of reactions of the form

$$X_1 + X_2 + \cdots + X_n \xrightarrow{\ k\ } Y_1 + Y_2 + \cdots + Y_m,$$

where X_i is a *reactant* and Y_j is a *product*. The positive real number k is the *rate constant* of the reaction, and along with the concentrations of the reactants, determines the speed at which the reaction progresses. Although reactions with any number of reactants and products are supported in the model, if they are restricted to exactly two, then stochastic chemical reaction networks are equivalent to population protocols [1].

Recent research has demonstrated that chemical reaction networks are capable of rich computation. They are Turing complete [13], and even weaker variants of the model such as rate-independent environments are still capable of computing a large class of functions [6,8,9]. The reachability problem in CRNs has also recently drawn the interest of computer scientists [5] as well as time complexity issues in the model [7,11]. Chemical reaction networks are also becoming increasingly practical, since they can be compiled into DNA strand displacement systems to be physically implemented [2].

Chemical reaction networks under deterministic mass action semantics are also related to the general purpose analog computer (GPAC), first introduced by Shannon [21]. Shannon's GPAC was inspired by MIT's differential analyzer, a machine designed to solve ordinary differential equations. The model has also recently been refined, and many remarkable results concerning its computational power have been published. The GPAC is Turing complete [17], is equivalent to systems of polynomial differential equations [16], and the class P of languages computable in polynomial time can be characterized with GPACs whose solutions have polynomial arc length [4]. Moreover, Fages, Le Guludec, Bournez, and

---

[1] The stochastic mass action model was proven to be "equivalent in the limit" to the deterministic mass action model as the number of molecules and the volume are scaled to infinity [20].

Pouly showed in 2017 that CRNs under deterministic mass action semantics can simulate GPACs by encoding each variable as the *difference* of two species concentrations [13].

In this paper, we investigate the relationship between real time computable real numbers by CRNs and general purpose analog computers. We define the class $\mathbb{R}_{RTGPAC}$ of real time computable real numbers by GPACs. Roughly, $\alpha \in \mathbb{R}_{RTGPAC}$ if there exists a polynomial initial value problem (PIVP) with integer coefficients such that, if initialized with all zeros, then all variables are bounded and one of the variables converges to α exponentially quickly. These restrictions are analogous to the definition of $\mathbb{R}_{RTCRN}$ and ensure that the PIVP is finitely specifiable and α is computed in real time. We show that $\mathbb{R}_{RTGPAC} = \mathbb{R}_{RTCRN}$ by proving that $\mathbb{R}_{RTCRN}$ is a subfield of $\mathbb{R}$ and using an extension of the difference representation introduced in [13] that relies on these closure properties. We also show that the constraint of all zero initial conditions can be relaxed to integral initial conditions. With these new theorems, we prove two well-known transcendental numbers e and π are members of $\mathbb{R}_{RTCRN}$. The proofs and constructions for these transcendental numbers are short and concise, and demonstrate the power of these theorems for generating and proving real-time CRNs correct.

The rest of this paper is organized as follows. Section 2 introduces relevant definitions and notations used throughout the paper; Sect. 3 includes the main theorem of the paper, that $\mathbb{R}_{RTGPAC} = \mathbb{R}_{RTCRN}$, along with the proof that $\mathbb{R}_{RTCRN}$ is a field; Sect. 4 includes proofs that e and π are real time computable by chemical reaction networks using the theorems from Sect. 3; and Sect. 5 provides some concluding remarks.

2   Preliminaries

Chemical reaction networks have been investigated from the perspective of chemistry [12], mathematics [14], and computer science [10], and each field uses slightly different notation. In this paper we use the notion introduced in [19], and thus define a *chemical reaction network* (*CRN*) to be an ordered pair $N = (S, R)$ where $S = \{Y_1, Y_2, \ldots, Y_n\}$ is a finite set of *species* and R is a finite set of *reactions*. A reaction is a triple $\rho = (\mathbf{r}, \mathbf{p}, k)$ where $\mathbf{r}, \mathbf{p} \in \mathbb{N}^n$ are vectors of reactant species and product species, respectively, and $k \in \mathbb{R}_{>0}$ is the rate constant. We use $\mathbf{r}_i$ and $\mathbf{p}_i$ to denote the ith component of $\mathbf{r}$ and $\mathbf{p}$, respectively. We call $Y_i \in S$ a *reactant* of ρ if $\mathbf{r}_i > 0$, a *product* of ρ if $\mathbf{p}_i > 0$, and a *catalyst* of ρ if $\mathbf{r}_i = \mathbf{p}_i > 0$.

In this paper, we are concerned with deterministic CRNs under mass action kinetics, and therefore the evolution of a CRN $N = (S, R)$ is governed by a system of real-valued functions $\mathbf{y} = (y_1, \ldots, y_n)$ where $y_i : \mathbb{R}_{\geq 0} \to \mathbb{R}_{\geq 0}$ for all $1 \leq i \leq n$. The value $y_i(t)$ is called the *concentration* of species $Y_i \in S$ at time $t \in [0, \infty)$. According to the law of mass action, the *rate* of a reaction $\rho = (\mathbf{r}, \mathbf{p}, k)$ is proportional to the rate constant k and the concentrations of the reactants $\mathbf{r}$.

Thus, the rate of a reaction ρ at time t is defined as

$$\text{rate}_\rho(t) = k \prod_{Y_i \in S} y_i(t)^{\mathbf{r}_i}. \tag{1}$$

The total rate of change of the concentration $y_i(t)$ of a species $Y_i \in S$ is the sum of the rates of reactions that affect Y_i, and thus is governed by the differential equation

$$y_i'(t) = \sum_{(\mathbf{r},\mathbf{p},k) \in R} (\mathbf{p}_i - \mathbf{r}_i) \cdot \text{rate}_\rho(t). \tag{2}$$

This system of ODEs, along with an initial condition $\mathbf{y}(0) \in \mathbb{R}^n_{>0}$, yields a *polynomial initial value problem (PIVP)* and has a unique solution $\mathbf{y}(t)$.

For convenience, we allow alternative species names such as X and Z and make use of more intuitive notation for specifying reactions. For example, a reaction $\rho = (\mathbf{r}, \mathbf{p}, k)$ can be written

$$X + Y \xrightarrow{\;\;k\;\;} X + 2Z, \tag{3}$$

where $\mathbf{r}$ and $\mathbf{p}$ are defined by the complexes $X + Y$ and $X + 2Z$, respectively. Thus X, Y are the reactants of ρ, X, Z are the products of ρ, and X is a catalyst of ρ. If a CRN $N = (S, R)$ consists of only the reaction above, this specifies the system of ODEs

$$x'(t) = 0,$$
$$y'(t) = -kx(t)y(t)$$
$$z'(t) = 2kx(t)y(t),$$

where $x(t)$, $y(t)$, and $z(t)$ are the concentrations of the species X, Y, and Z at time t. In this paper we use this intuitive notation whenever possible.

Two CRNs can also naturally be combined into one. Given $N_1 = (S_1, R_1)$ and $N_2 = (S_2, R_2)$, we define the *join* of N_1 and N_2 to be the CRN

$$N_1 \sqcup N_2 = (S_1 \cup S_2, R_1 \cup R_2). \tag{4}$$

For a CRN $N = (S, R)$ and a species $X \in S$, we say that (N, X) is a *computer* for a real number α if the following three properties hold:

1. **Integral Rate Constants.** For each reaction $\rho = (\mathbf{r}, \mathbf{p}, k) \in R$, the rate constant $k \in \mathbb{Z}_{>0}$ is a positive integer.
2. **Bounded Concentrations.** There is a constant β such that, if all species concentrations are initialized to 0, then $y_i(t) \leq \beta$ for each species $Y_i \in S$ and for all time $t \geq 0$.
3. **Real-Time Convergence.** If all species in N are initialized to 0 at time 0, then for all $t \geq 1$,

$$|x(t) - |\alpha|| \leq 2^{-t}. \tag{5}$$

The real numbers for which there is a computer (N, X) are called *real-time CRN computable*. The set of real-time CRN computable real numbers is denoted by $\mathbb{R}_{RTCRN}$.

The restriction to integral rate constants ensures that the CRN can be specified in finitely many bits. Bounded concentrations imposes a limit on the rate of convergence, which makes the variable t a meaningful measure of time. If concentrations were unbounded, it is possible to compute a real number arbitrarily quickly by catalyzing all reactions with a species whose concentration is unbounded. Bournez, Graça, and Pouly recently showed that the *arc length* of $\mathbf{y}(t)$ is a better time complexity metric [4], and bounded concentrations immediately implies the arc length is linear in t, making them equivalent metrics in this case. The real-time convergence requires that α is computed with $\lfloor t \rfloor$ bits of accuracy in t units of time, making it analogous to the definition of real time Turing computation.

The definition of $\mathbb{R}_{RTCRN}$ can also be generalized in the following way. Given two real numbers $\tau, \gamma \in (0, \infty)$, we define $\mathbb{R}_{RTCRN}^{\tau, \gamma}$ in the same way as $\mathbb{R}_{RTCRN}$ except the real-time convergence constraint (5) is replaced with:

$$|x(t) - |\alpha|| \le e^{-\gamma t}, \tag{6}$$

for all $t \ge \tau$. Proven in [19] but restated here for convenience is the following lemma:

Lemma 1. $\mathbb{R}_{RTCRN} = \mathbb{R}_{RTCRN}^{\tau, \gamma}$ *for all* $\tau, \gamma \in (0, \infty)$.

The above lemma shows that the definition of $\mathbb{R}_{RTCRN}$ is robust to linear changes to the real-time convergence rate. Thus, it suffices to show that a CRN computes a real number α exponentially quickly, to prove that $\alpha \in \mathbb{R}_{RTCRN}$ without explicitly showing that $\tau = 1$ and $\gamma = \ln 2$.

3   Real-Time Equivalence of CRNs and GPACs

This section is devoted to proving that the class $\mathbb{R}_{RTCRN}$ is equivalent to an analogous class $\mathbb{R}_{RTGPAC}$ of real time computable real numbers by general purpose analog computers. We begin by formally defining $\mathbb{R}_{RTGPAC}$.

For a PIVP $\mathbf{y} = (y_1, y_2, \ldots, y_n)$ satisfying $\mathbf{y}(0) = \mathbf{0}$, we say that $\mathbf{y}$ is an *computer* for a real number α if the following three properties hold:

1. All coefficients of $\mathbf{y}$ are integers,
2. There is a constant $\beta > 0$ such that $|y_i(t)| \le \beta$ for all $1 \le i \le n$ and $t \in [0, \infty)$, and
3. $|y_1(t) - \alpha| \le 2^{-t}$ for all $t \in [1, \infty)$.

The real numbers for which there is a computer $\mathbf{y}$ are called *real-time GPAC computable*. The set of real-time CRN computable real numbers is denoted by $\mathbb{R}_{RTGPAC}$.

Note that the constraints above mirror the definition of $\mathbb{R}_{RTCRN}$ in Sect. 2 except for the fact that $y_1(t)$ is converging to α instead of $|\alpha|$. This difference is due to the CRN restriction of species concentrations to be non-negative real numbers whereas the value of a GPAC variable $y_i(t)$ has no such restriction.

Lemma 2. $\mathbb{R}_{RTCRN} \subseteq \mathbb{R}_{RTGPAC}$.

Proof. Given a computer (N, Y_1) for $\alpha \in \mathbb{R}$, let $\mathbf{y}$ be the PIVP induced by the deterministic semantics of N from Eq. (2). Note that (N, Y_1) computes α when its species concentrations are initialized to zero, therefore $\mathbf{y}(0) = \mathbf{0}$. The fact that $\mathbf{y}$ is also a computer for α immediately follows from the constraints imposed on N and the fact that if $\alpha < 0$, a multiplying each ODE by -1 causes $y_1(t)$ to converge directly to α instead of $|\alpha|$.   □

Although the inclusion above is trivial, the fact that $\mathbb{R}_{RTGPAC} \subseteq \mathbb{R}_{RTCRN}$ is not so obvious. This is due to deterministic CRNs inducing PIVPs with restricted forms, namely, the polynomial of each ODE has the structure

$$y'(t) = p(t) - q(t)y(t),$$

where p and q are polynomials over the concentrations of the species. The fact that negative terms in the ODE for Y must depend on its own concentration $y(t)$ makes certain GPAC constructions difficult to implement with CRNs.

The rest of this section is devoted to finishing the proof of the main theorem: $\mathbb{R}_{RTGPAC} = \mathbb{R}_{RTCRN}$. To simplify the proof, we first prove that $\mathbb{R}_{RTCRN}$ is a subfield of $\mathbb{R}$ which solves an open problem stated in [19]. The proofs of closure under addition, multiplication, division, and subtraction rely on certain convergence properties. Thus, we first state and prove two lemmas which demonstrate that certain differential equations immediately yield exponential convergence to a target real number. Then we prove the four closure properties necessary to show that $\mathbb{R}_{RTCRN}$ is a field using these lemmas. Finally, we conclude with the proof of the main theorem that $\mathbb{R}_{RTGPAC} = \mathbb{R}_{RTCRN}$.

Lemma 3 (Direct Convergence Lemma). *If $\alpha \in \mathbb{R}$ and $x, f : [0, \infty) \to \mathbb{R}$ are functions that satisfy*

$$x'(t) = f(t) - x(t) \text{ for all } t \in [0, \infty) \tag{7}$$

$$|f(t) - \alpha| \le e^{-t} \text{ for all } t \in [1, \infty), \tag{8}$$

then there exist constants $\gamma, \tau \in (0, \infty)$ such that

$$|x(t) - \alpha| \le e^{-\gamma t} \text{ for all } t \in [\tau, \infty). \tag{9}$$

Proof. Assume the hypothesis. The ODE of Eq. (7) can be solved directly using the integrating factor method and has a solution of the form

$$x(t) = e^{-t} \int e^t f(t) dt. \tag{10}$$

By Eq. (8), we know that for all $t \geq 1$,

$$\int e^t f(t) dt \leq \int e^t \left(\alpha + e^{-t} \right) dt = \alpha e^t + t + C_1,$$

for some constant C_1. This, along with Eq. (10), yields

$$x(t) \leq \alpha + e^{-t} \left(t + C_1 \right). \tag{11}$$

Using a similar argument, it is easy to show that

$$x(t) \geq \alpha - e^{-t} \left(t + C_2 \right) \tag{12}$$

for some constant C_2. Choosing $C = \max \{ 0, C_1, C_2 \}$, it follows from Eqs. (11) and (12) that

$$|x(t) - \alpha| \leq (t + C) e^{-t} \leq e^{-t/2},$$

for all $t \geq \max\{1, 4\log(C + 1)\}$. $\square$

Lemma 4 (Reciprocal Convergence Lemma). *If $\alpha \in \mathbb{R}_{>0}$ and $x, f : [0, \infty) \to \mathbb{R}$ are continuous functions that satisfy*

$$x'(t) = 1 - f(t) \cdot x(t) \text{ for all } t \in [0, \infty) \tag{13}$$

$$|f(t) - \alpha| \leq e^{-t} \text{ for all } t \in [1, \infty), \tag{14}$$

then there exist constants $\gamma, \tau > 0$ such that

$$\left| x - \frac{1}{\alpha} \right| \leq e^{-\gamma t} \text{ for all } t \in [\tau, \infty). \tag{15}$$

Proof. Assume the hypothesis. Since f is continuous, its antiderivative exists, so the ODE from Eq. (13) can be solved directly using the integrating factor method with a solution of the form

$$x(t) = e^{-F(t)} \int_0^t e^{F(s)} ds, \tag{16}$$

where $F(t) = \int_0^t f(s) ds$. If we let $h(t) = f(t) - \alpha$, and let $H(t) = \int_0^t h(s) ds$ be the antiderivative of h, then

$$F(t) = \int_0^t \left(\alpha + h(s) \right) ds = \alpha t + H(t).$$

Using this relationship, we can rewrite Eq. (16) as

$$x(t) = e^{-F(t)} \cdot \frac{1}{\alpha} \int_0^t e^{H(s)} \left(\alpha e^{\alpha s} \right) ds. \tag{17}$$

We can now use integration by parts on the integral of Eq. (17) with $u(s) = e^{H(s)}$ and $v'(s) = \alpha e^{\alpha s}$ to obtain

$$\int_0^t e^{H(s)} \left(\alpha e^{\alpha s}\right) ds = \int_0^t u(s)v'(s)ds$$

$$= u(s)v(s)\Big|_0^t - \int_0^t v(s)u'(s)ds$$

$$= e^{H(t)}e^{\alpha t} - 1 - \int_0^t e^{\alpha s} \left(h(s)e^{H(s)}\right) ds.$$

Substituting this into Eq. (17) and using the fact that $F(t) = \alpha t + H(t)$, we obtain

$$x(t) = e^{-F(t)} \cdot \frac{1}{\alpha} \left(e^{F(t)} - 1 - \int_0^t h(s)e^{F(s)}ds\right),$$

which yields the following bound:

$$\left|x(t) - \frac{1}{\alpha}\right| \le e^{-F(t)} \left(1 + \int_0^t |h(s)|e^{F(s)}ds\right). \tag{18}$$

It remains to be shown that the right-hand side of Eq. (18) is bounded by an exponential after some time τ. We begin by showing that $H(t)$ is bounded above and below by the constant $C_1 = \int_0^1 |h(s)|ds + \frac{1}{e}$:

$$|H(t)| \le \int_0^t |h(s)|ds \le \int_0^1 |h(s)|ds + \int_1^t e^{-s}ds = C_1 - e^{-t} \le C_1.$$

It immediately follows that

$$e^{F(t)} = e^{\alpha t + H(t)} \le C_2 e^{\alpha t}$$

$$e^{-F(t)} = e^{-\alpha t - H(t)} \le C_2 e^{-\alpha t}$$

where $C_2 = e^{C_1}$. If we define the constant $C_3 = \int_0^1 |h(s)|e^{F(s)}ds$, we can bound the integral of Eq. (18) with

$$\int_0^t |h(s)|e^{F(s)}ds \le C_3 + \int_1^t e^{-s} \left(C_2 e^{\alpha s}\right) ds = C_5 + C_4 e^{(\alpha-1)t}$$

where $C_4 = \frac{C_2}{\alpha-1}$ and $C_5 = C_3 - C_4 e^{\alpha-1}$. Thus, we can rewrite Eq. (18):

$$\left|x(t) - \frac{1}{\alpha}\right| \le C_2 e^{-\alpha t} \left(1 + C_5 + C_4 e^{(\alpha-1)t}\right) = C_6 e^{-\alpha t} + C_7 e^{-t}$$

where $C_6 = C_2(1 + C_5)$ and $C_7 = C_2 C_4$.

It immediately follows that there exist constants γ and τ such that $\left|x(t) - \frac{1}{a}\right|$ is bounded by $e^{-\gamma t}$ for all $t \in [\tau, \infty)$. □

Using Lemmas 3 and 4, we now prove that $\mathbb{R}_{RTCRN}$ is a field. We split the four closure properties into the following four lemmas.

Lemma 5. *If* $\alpha, \beta \in \mathbb{R}_{RTCRN}$, *then* $\alpha + \beta \in \mathbb{R}_{RTCRN}$.

Proof. Assume the hypothesis, and let (N_α, X) and (N_β, Y) be CRN computers that compute α and β, respectively. Without loss of generality, we assume that $\alpha, \beta \geq 0$ and that N_α and N_β do not share any species.

Now let Z be a new species, and let $N = N_\alpha \sqcup N_\beta \sqcup \widehat{N}$ where $\widehat{N}$ is the CRN defined by the reactions

$$X \xrightarrow{1} X + Z$$
$$Y \xrightarrow{1} Y + Z$$
$$Z \xrightarrow{1} \emptyset.$$

Note that the species in N_α and N_β are unaffected by the reactions of $\widehat{N}$, and the ODE for Z is:

$$z'(t) = x(t) + y(t) - z(t). \tag{19}$$

Let $f(t) = x(t) + y(t)$. By Lemma 1, without loss of generality, we can assume that $|f(t) - \alpha - \beta| \leq e^{-t}$ for all $t \geq 1$. Immediately by Lemmas 3 and 1, we conclude that $\alpha + \beta \in \mathbb{R}_{RTCRN}$.   □

Lemma 6. *If* $\alpha, \beta \in \mathbb{R}_{RTCRN}$, *then* $\alpha\beta \in \mathbb{R}_{RTCRN}$.

Proof. Assume the hypothesis, and let (N_α, X) and (N_β, Y) be CRN computers that compute α and β, respectively Furthermore, we assume that N_α and N_β do not share any species. Without loss of generality, we also assume that $\alpha, \beta \geq 0$.

Now let Z be a new species, and let $N = N_\alpha \sqcup N_\beta \sqcup \widehat{N}$ where $\widehat{N}$ is the CRN defined by the reactions

$$X + Y \xrightarrow{1} X + Y + Z$$
$$Z \xrightarrow{1} \emptyset.$$

Note that the species in N_α and N_β are unaffected by the reactions of $\widehat{N}$ and yields the following ODE for Z:

$$z'(t) = x(t)y(t) - z(t). \tag{20}$$

Let $f(t) = x(t)y(t)$. By Lemma 1, without out loss of generality, we can assume that $|f(t) - \alpha\beta| \leq e^{-t}$ for all $t \geq 1$. Immediately by Lemmas 3 and 1, we conclude that $\alpha\beta \in \mathbb{R}_{RTCRN}$.   □

Lemma 7. *If* $\alpha \in \mathbb{R}_{RTCRN}$ *and* $\alpha \neq 0$, *then* $\frac{1}{\alpha} \in \mathbb{R}_{RTCRN}$.

Proof. Assume the hypothesis, and let (N_α, X) be CRN a computer that testifies to this. Without loss of generality, we also assume that $\alpha > 0$.

Now let Y be a new species, and let $N = N_\alpha \sqcup \widehat{N}$ where $\widehat{N}$ is the CRN defined by the reactions

$$\emptyset \xrightarrow{1} Y$$

$$X + Y \xrightarrow{1} X.$$

Note that the species in N_α are unaffected by the reactions of $\widehat{N}$ and yields the following ODE for Y:

$$z'(t) = 1 - x(t)y(t). \tag{21}$$

Since $\alpha \in \mathbb{R}_{RTCRN}$, we know that $|f(t) - \alpha| \le e^{-t}$ for all $t \ge 1$. It follows from Lemmas 4 and 1 that $\frac{1}{\alpha} \in \mathbb{R}_{RTCRN}$.   □

Lemma 8. *If* $\alpha, \beta \in \mathbb{R}_{RTCRN}$, *then* $\alpha - \beta \in \mathbb{R}_{RTCRN}$.

Proof. Assume the hypothesis, and let (N_α, X) and (N_β, Y) be CRN computers that compute α and β, respectively. Furthermore, we assume that N_α and N_β do not share any species. Without loss of generality, we also assume that $\alpha > \beta \ge 0$.

Now let Z be a new species, and let $N = N_\alpha \sqcup N_\beta \sqcup \widehat{N}$ where $\widehat{N}$ is the CRN defined by the reactions

$$\emptyset \xrightarrow{1} Z$$

$$X + Z \xrightarrow{1} X$$

$$Y + Z \xrightarrow{1} Y + 2Z.$$

Note that the species in N_α and N_β are unaffected by the reactions of $\widehat{N}$ and yields the following ODE for Z:

$$z'(t) = 1 - (x(t) - y(t))z(t). \tag{22}$$

Let $f(t) = x(t) - y(t)$. By Lemma 1, without out loss of generality, we can assume that $|f(t) - (\alpha - \beta)| \le e^{-t}$ for all $t \ge 1$. By Lemmas 4 and 1, we know that $\frac{1}{\alpha - \beta} \in \mathbb{R}_{RTCRN}$. By Lemmas 7, we conclude that $\alpha - \beta \in \mathbb{R}_{RTCRN}$.   □

Theorem 1. $\mathbb{R}_{RTCRN}$ *is a subfield of* $\mathbb{R}$.

Proof. This immediately follows from Lemmas 5–8 and the fact that $\mathbb{R}_{RTCRN}$ is non-empty.   □

As a consequence of Theorem 1, and the results of [19] we now know that $\mathbb{R}_{RTCRN}$ contains all algebraic numbers and an infinite family of transcendental numbers. However, we have yet to prove that natural transcendentals such as e and π are real-time computable by CRNs. These proofs are simplified dramatically using the following theorem which uses a construction similar to [13].

Theorem 2. $\mathbb{R}_{RTCRN} = \mathbb{R}_{RTGPAC}$.

Proof. We have already shown the forward direction in Lemma 2.

For the backward direction, assume that $0 \neq \alpha \in \mathbb{R}_{RTGPAC}$, and let $\mathbf{y} = (y_1, y_2, \ldots, y_n)$ be the PIVP that testifies to this. Then the individual components of $\mathbf{y}$ obey the ODEs

$$y_1' = p_1(y_1, \ldots, y_n),$$
$$y_2' = p_2(y_1, \ldots, y_n),$$
$$\vdots$$
$$y_n' = p_n(y_1, \ldots, y_n).$$

For each $1 \leq i \leq n$, we define the variables $\hat{\mathbf{y}} = (z, u_1, v_1, u_2, v_2, \ldots, u_n, v_n)$ as well as the polynomials

$$\hat{p}_i(\hat{\mathbf{y}}) = p_i(u_1 - v_1, u_2 - v_2, \ldots, u_n - v_n),$$

noting that each $\hat{p}_i$ is indeed an integral polynomial over the variables of $\hat{\mathbf{y}}$. For each $1 \leq i \leq n$, we also define the polynomials $\hat{p}_i^+$ and $\hat{p}_i^-$ by the positive and negative terms of $\hat{p}_i$, respectively, whence $\hat{p}_i = \hat{p}_i^+ - \hat{p}_i^-$.

We now define ODEs for each variable u_i and v_i of $\hat{\mathbf{y}}$,

$$u_i' = \hat{p}_i^+ - u_i v_i \left(\hat{p}_i^+ + \hat{p}_i^- \right), \tag{23}$$
$$v_i' = \hat{p}_i^- - u_i v_i \left(\hat{p}_i^+ + \hat{p}_i^- \right), \tag{24}$$

as well as the ODE for the variable z

$$z' = 1 - (u_1 - v_1)z. \tag{25}$$

Notice that if $y_i = u_i - v_i$, then

$$u_i' - v_i' = \hat{p}_i^+ - \hat{p}_i^- = \hat{p}_i = p_i = y_i',$$

therefore if $\hat{\mathbf{y}}(0) = \mathbf{0}$, we know that $y_i(t) = u_i(t) - v_i(t)$ for all $t \in [0, \infty)$.

We now prove that every variable of $\hat{\mathbf{y}}$ is bounded from above by some constant. For the sake of contradiction, assume that either u_i or v_i is unbounded. Recall that each variable of $\mathbf{y}$ is bounded by some $\beta > 0$, and therefore $-\beta \leq y_i(t) \leq \beta$ for all $t \in [0, \infty)$. Since $y_i(t) = u_i(t) - v_i(t)$, it follows that *both* u_i and v_i must be unbounded. However, this is a contradiction since u_i' and v_i' each include the negative terms $-u_i v_i \left(\hat{p}_i^+ + \hat{p}_i^- \right)$ which grow faster than their positive terms. Thus, u_i and v_i must both be bounded.

Since each of the ODEs of $\hat{\mathbf{y}}$ can be written in the form $x' = p - qx$ where p and q are polynomials with positive integral coefficients, there exists a CRN $N = (S, R)$ with species $S = \{U_i, V_i \mid 1 \leq i \leq n\} \cup \{Z\}$ that obey these ODEs. Because $y_1 = u_1 - v_1$, this means that $|u_1(t) - v_1(t) - \alpha| \leq 2^{-t}$. By Lemma 4, it immediately follows that N real time computes $\frac{1}{\alpha}$ with species Z. Finally, we obtain that $\alpha \in \mathbb{R}_{RTCRN}$ by closure under reciprocal. $\square$

4 e and π Are Real-Time Computable by CRNs

In this section, we will prove that e and π are real time computable by CRNs, which was not previously known. However, first we prove a useful theorem that shows that the constraint that the CRN or GPAC must be initialized to all zeros can be relaxed to any integral initial condition. This theorem dramatically simplifies the constructions, since the numbers e and π can be naturally computed if a species is initialized to 1.

Theorem 3. *If $\alpha \in \mathbb{R}$ and $\mathbf{y} = (y_1, y_2, \ldots, y_n)$, $\mathbf{y}(0) \in \mathbb{Z}^n$ is a PIVP such that*

1. *$|y_i(t)| \leq \beta$ for all $1 \leq i \leq n$ and $t \in [0, \infty)$ for some $\beta > 0$, and*
2. *$|y_1(t) - \alpha| \leq 2^{-t}$ for all $t \in [0, \infty)$,*

then $\alpha \in \mathbb{R}_{RTGPAC}$.

Proof. Assume the hypothesis. Then there is a polynomial p_i corresponding to each variable y_i of $\mathbf{y}$ such that $y_i' = p_i$. We will now define a related PIVP that when initialized to all zeros computes α.

Define the variables $\hat{\mathbf{y}} = (\hat{y}_1, \hat{y}_2, \ldots, \hat{y}_n)$ that obey the ODEs

$$\hat{y}_i' = p_i(\hat{y}_1 + y_1(0), \hat{y}_2 + y_2(0), \ldots, \hat{y}_n + y_n(0)).$$

Since $\mathbf{y}(0) \in \mathbb{Z}^n$, each ODE $\hat{y}_i$ is a polynomial with integral coefficients. We also note that if $\hat{y}_i(t) = y_i(t) - y_i(0)$ for some $t \in [0, \infty)$, then

$$\hat{y}_i'(t) = p_i(y_1(t), y_2(t), \ldots, y_n(t)) = y_i'(t).$$

Thus, if we initialize $\hat{\mathbf{y}}(0) = \mathbf{0}$, it follows that $\hat{y}_i(t) = y_i(t) - y_i(0)$ for all $t \in [0, \infty)$. Since the PIVP $\mathbf{y}$ computes α, it follows that the PIVP $\hat{\mathbf{y}}$ computes $\alpha - y_1(0)$, and therefore $\alpha - y_1(0) \in \mathbb{R}_{RTGPAC}$.

Finally, since $y_1(0) \in \mathbb{Z}$, it is also in $\mathbb{R}_{RTGPAC}$, and by closure under addition we conclude that $\alpha \in \mathbb{R}_{RTGPAC}$.   □

We now present concise proofs that the e and π are members of $\mathbb{R}_{RTCRN}$.

Theorem 4. $e \in \mathbb{R}_{RTCRN}$.

Proof. By Theorem 3, it suffices to show that there exists a CRN computer with integral initial conditions that computes e exponentially quickly. Consider the CRN defined by

$$X \xrightarrow{1} \emptyset$$

$$X + Y \xrightarrow{1} X$$

along with the initial condition $x(0) = 1$ and $y(0) = 1$. This induces the system of ODES

$$x'(t) = -x(t) \tag{26}$$
$$y'(t) = -x(t)y(t), \tag{27}$$

which is trivial to solve and has solution

$$x(t) = e^{-t}, \quad y(t) = e^{1-e^{-t}}.$$

It is clear that $y(t)$ exponentially goes to e, and thus $e \in \mathbb{R}_{RTCRN}$.   □

It is easy to apply the construction of Theorem 3 to the CRN provided in the proof of Theorem 4, and Fig. 1 shows the plot of this expanded CRN computing e in this way.

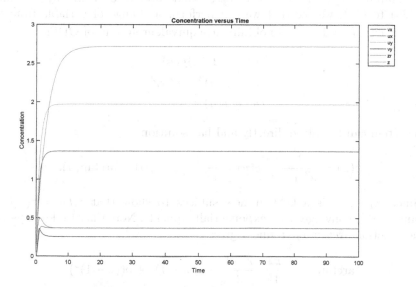

Fig. 1. MATLAB visualization of computing e from Theorem 4. This plot is of the CRN after applying the construction from Theorem 3 so that all species concentrations are initially zero, and Z is the species converging to e.

Theorem 5. $\pi \in \mathbb{R}_{RTCRN}$.

Proof. By Theorem 3, it suffices to show that there exists a CRN computer with integral initial conditions that computes π exponentially quickly. Consider the CRN defined by

$$W \xrightarrow{1} \emptyset$$

$$W + X + Y \xrightarrow{2} W + Y$$

$$W + 2X \xrightarrow{1} W + 2X + Y$$

$$W + 2Y \xrightarrow{1} W + Y$$

$$W + X \xrightarrow{1} W + X + Z,$$

with initial condition $w(0) = x(0) = 1$ and $y(0) = z(0) = 0$. It is easy to verify that this CRN induces the following system of ODEs

$$w'(t) = -w(t), \tag{28}$$
$$x'(t) = -2w(t)x(t)y(t), \tag{29}$$
$$y'(t) = w(t)x(t)^2 - w(t)y(t)^2, \tag{30}$$
$$z'(t) = w(t)x(t). \tag{31}$$

By examining Eq. (28), it is easy to see that $w(t) = e^{-t}$, and by examining Eqs. (29) to (31), we see that we can perform a change of variable from t to $u(t) = \int_0^t w(s)ds = 1 - e^{-t}$ to obtain the equivalent system of ODEs:

$$x'(u) = -2x(u)y(u),$$
$$y'(u) = x(u)^2 - y(u)^2,$$
$$z'(u) = x(u).$$

This system can be solved directly and has solution

$$x(u) = \frac{1}{u^2 + 1}, \quad y(u) = \frac{u}{u^2 + 1}, \quad z(u) = \arctan(u).$$

Since $\mathbb{R}_{RTCRN}$ is a field, it now suffices to show that $z(t) = z(u(t)) = \arctan(1 - e^{-t})$ converges to $\frac{\pi}{4}$ exponentially quickly. Note that Taylor expansion of the function $\arctan(x)$ around 1 gives

$$\arctan(x) = \frac{\pi}{4} + \frac{x-1}{2} - \frac{1}{4}(x-1)^2 + o((x-1)^2).$$

Thus we obtain

$$\arctan(u(t)) - \frac{\pi}{4} = O(u(t) - 1) = O(e^{-t}).$$

Hence $\arctan(1 - e^{-t})$ converges to $\frac{\pi}{4}$ exponentially quickly, and therefore $\pi \in \mathbb{R}_{RTCRN}$. $\qquad \square$

It is easy to generate the reactions of the explicit CRN that computes π from an all-zero initial condition. The plot of this CRN is provided in Fig. 2.

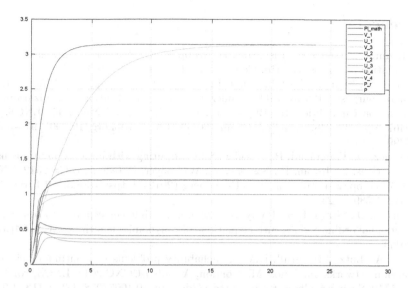

Fig. 2. MATLAB visualization of computing π from Theorem 4. This plot is of the CRN after applying the construction from Theorem 3 so that all species concentrations are initially zero, and P is the species converging to π.

5   Conclusion

In this paper, we investigated the relationship of the class $\mathbb{R}_{RTCRN}$ of real time computable real numbers with chemical reaction networks and the class $\mathbb{R}_{RTGPAC}$ of real time computable real numbers with general purpose analog computers. In particular, we proved that $\mathbb{R}_{RTGPAC} = \mathbb{R}_{RTCRN}$ by first solving the previously open problem posed in [19] that $\mathbb{R}_{RTCRN}$ is a field, and then simulating the GPAC with a CRN in a similar way to [13]. In particular, we extend their construction so that our CRN simulation computes the real number α with a *single* species instead of the difference of two species concentrations. We prove this using the reciprocal convergence lemma and the fact that $\mathbb{R}_{RTCRN}$ is closed under reciprocal.

We also used the GPAC equivalence to prove that the restriction to all zero initial conditions is not necessary and can be relaxed to integral initial conditions. This led to concise and natural proofs that e and π are in $\mathbb{R}_{RTCRN}$. We should note that applying the constructions in the proofs of Theorems 2 and 3 leads to CRNs with a large number of reactions.

We hope future research will uncover techniques to compute numbers such as e and π more efficiently, as well as research that leads to a better understanding of the structure of $\mathbb{R}_{RTCRN}$.

Acknowledgments. We thank Jack Lutz for helpful comments and suggestions. We also thank the anonymous reviews for their input, and especially insightful comments concerning the presentation of this paper.

References

1. Angluin, D., Aspnes, J., Eisenstat, D., Ruppert, E.: The computational power of population protocols. Distrib. Comput. **20**(4), 279–304 (2007)
2. Badelt, S., Shin, S.W., Johnson, R.F., Dong, Q., Thachuk, C., Winfree, E.: A general-purpose CRN-to-DSD compiler with formal verification, optimization, and simulation capabilities. In: Brijder, R., Qian, L. (eds.) DNA 2017. LNCS, vol. 10467, pp. 232–248. Springer, Cham (2017). https://doi.org/10.1007/978-3-319-66799-7_15
3. Bournez, O., Fraigniaud, P., Koegler, X.: Computing with large populations using interactions. In: Rovan, B., Sassone, V., Widmayer, P. (eds.) MFCS 2012. LNCS, vol. 7464, pp. 234–246. Springer, Heidelberg (2012). https://doi.org/10.1007/978-3-642-32589-2_23
4. Bournez, O., Graça, D.S., Pouly, A.: Polynomial time corresponds to solutions of polynomial ordinary differential equations of polynomial length. J. ACM **64**(6), 38 (2017)
5. Case, A., Lutz, J.H., Stull, D.M.: Reachability problems for continuous chemical reaction networks. In: Amos, M., Condon, A. (eds.) UCNC 2016. LNCS, vol. 9726, pp. 1–10. Springer, Cham (2016). https://doi.org/10.1007/978-3-319-41312-9_1
6. Chalk, C., Kornerup, N., Reeves, W., Soloveichik, D.: Composable rate-independent computation in continuous chemical reaction networks. In: Češka, M., Šafránek, D. (eds.) CMSB 2018. LNCS, vol. 11095, pp. 256–273. Springer, Cham (2018). https://doi.org/10.1007/978-3-319-99429-1_15
7. Chen, H.L., Cummings, R., Doty, D., Soloveichik, D.: Speed faults in computation by chemical reaction networks. Distrib. Comput. **30**(5), 373–390 (2017)
8. Chen, H.-L., Doty, D., Soloveichik, D.: Deterministic function computation with chemical reaction networks. In: Stefanovic, D., Turberfield, A. (eds.) DNA 2012. LNCS, vol. 7433, pp. 25–42. Springer, Heidelberg (2012). https://doi.org/10.1007/978-3-642-32208-2_3
9. Chen, H.L., Doty, D., Soloveichik, D.: Rate-independent computation in continuous chemical reaction networks. In: Proceedings of the 5th Conference on Innovations in Theoretical Computer Science, pp. 313–326. ACM (2014)
10. Cook, M., Soloveichik, D., Winfree, E., Bruck, J.: Programmability of chemical reaction networks. In: Condon, A., Harel, D., Kok, J.N., Salomaa, A., Winfree, E. (eds.) Algorithmic Bioprocesses. Natural Computing Series, pp. 543–584. Springer, Heidelberg (2009). https://doi.org/10.1007/978-3-540-88869-7_27
11. Doty, D.: Timing in chemical reaction networks. In: Proceedings of the 25th Symposium on Discrete Algorithms, pp. 772–784 (2014)
12. Epstein, I.R., Pojman, J.A.: An Introduction to Nonlinear Chemical Dynamics: Oscillations, Waves, Patterns, and Chaos. Oxford University Press, Oxford (1998)
13. Fages, F., Le Guludec, G., Bournez, O., Pouly, A.: Strong turing completeness of continuous chemical reaction networks and compilation of mixed analog-digital programs. In: Feret, J., Koeppl, H. (eds.) CMSB 2017. LNCS, vol. 10545, pp. 108–127. Springer, Cham (2017). https://doi.org/10.1007/978-3-319-67471-1_7
14. Feinberg, M.: Foundations of Chemical Reaction Network Theory. AMS, vol. 202. Springer, Cham (2019). https://doi.org/10.1007/978-3-030-03858-8
15. Fischer, P.C., Meyer, A.R., Rosenberg, A.L.: Time-restricted sequence generation. J. Comput. Syst. Sci. **4**(1), 50–73 (1970)
16. Graça, D.S.: Some recent developments on shannon's general purpose analog computer. Math. Logic Q.: Math. Logic Q. **50**(4–5), 473–485 (2004)

17. Graça, D.S., Costa, J.F.: Analog computers and recursive functions over the reals. J. Complex. **19**(5), 644–664 (2003)
18. Hartmanis, J., Stearns, R.E.: On the computational complexity of algorithms. Trans. Am. Math. Soc. **117**, 285–306 (1965)
19. Huang, X., Klinge, T.H., Lathrop, J.I., Li, X., Lutz, J.H.: Real-time computability of real numbers by chemical reaction networks. Nat. Comput. **18**(1), 63–73 (2019)
20. Kurtz, T.G.: The relationship between stochastic and deterministic models for chemical reactions. J. Chem. Phys. **57**(7), 2976–2978 (1972)
21. Shannon, C.E.: Mathematical theory of the differential analyzer. Stud. Appl. Math. **20**(1–4), 337–354 (1941)
22. Yamada, H.: Real-time computation and recursive functions not real-time computable. IRE Trans. Electron. Comput. **EC–11**(6), 753–760 (1962)

A Reaction Network Scheme
Which Implements Inference
and Learning for Hidden Markov Models

Abhinav Singh[1]([✉]), Carsten Wiuf[2], Abhishek Behera[3],
and Manoj Gopalkrishnan[3]

[1] UM-DAE Centre for Excellence in Basic Sciences, Mumbai, India
abhinavsns7@gmail.com
[2] Department of Mathematical Sciences, University of Copenhagen,
Copenhagen, Denmark
wiuf@math.ku.dk
[3] Indian Institute of Technology Bombay, Mumbai, India
abhishek.enlightened@gmail.com, manoj.gopalkrishnan@gmail.com

Abstract. With a view towards molecular communication systems and molecular multi-agent systems, we propose the Chemical Baum-Welch Algorithm, a novel reaction network scheme that learns parameters for Hidden Markov Models (HMMs). Each reaction in our scheme changes only one molecule of one species to one molecule of another. The reverse change is also accessible but via a different set of enzymes, in a design reminiscent of futile cycles in biochemical pathways. We show that every fixed point of the Baum-Welch algorithm for HMMs is a fixed point of our reaction network scheme, and every positive fixed point of our scheme is a fixed point of the Baum-Welch algorithm. We prove that the "Expectation" step and the "Maximization" step of our reaction network separately converge exponentially fast. We simulate mass-action kinetics for our network on an example sequence, and show that it learns the same parameters for the HMM as the Baum-Welch algorithm.

1 Introduction

The sophisticated behavior of living cells on short timescales is powered by biochemical reaction networks. One may say that evolution has composed the symphony of the biosphere, genetic machinery conducts the music, and reaction networks are the orchestra. Understanding the capabilities and limits of this molecular orchestra is key to understanding living systems, as well as to engineering molecular systems that are capable of sophisticated life-like behavior.

The technology of implementing abstract reaction networks with molecules is a subfield of molecular systems engineering that has witnessed rapid advances in recent times. Several researchers [1–8] have proposed theoretical schemes for implementing arbitrary reaction networks with DNA oligonucleotides. There is a growing body of experimental demonstrations of such schemes [2,7,9–11].

© Springer Nature Switzerland AG 2019
C. Thachuk and Y. Liu (Eds.): DNA 25, LNCS 11648, pp. 54–79, 2019.
https://doi.org/10.1007/978-3-030-26807-7_4

A stack of tools is emerging to help automate the design process. We can now compile abstract reaction networks to a set of DNA oligonucleotides that will implement the dynamics of the network in solution [12]. We can computationally simulate the dynamics of these oligonucleotide molecular systems [13] to allow debugging prior to experimental implementation. In view of these rapid advances, the study of reaction networks from the point of view of their computational capabilities has become even more urgent.

It has long been known that reaction networks can compute any computable function [14]. The literature has several examples of reaction network schemes that have been inspired by known algorithms [15–22]. Our group has previously described reaction network schemes that solve statistical problems like maximum likelihood [23], sampling a conditional distribution and inference [24], and learning from partial observations [25]. These schemes exploit the thermodynamic nature of the underlying molecular systems that will implement these reaction networks, and can be expressed in terms of variational ideas involving minimization of Helmholtz free energy [26–28].

In this paper, we consider situations where partial information about the environment is available to a cell in the form of a sequence of observations. For example, this might happen when an enzyme is acting processively on a polymer, or a molecular walker [29–31] is trying to locate its position on a grid. In situations like this, multiple observations are not independent. Such sequences can not be summarized merely by the *type* of the sequence [32], i.e., the number of times various symbols occur. Instead, the order of various observations carries information about state changes in the process producing the sequence. The number of sequences grows exponentially with length, and our previously proposed schemes are algorithmically inadequate. To deal with such situations requires a pithy representation of sequences, and a way of doing inference and learning directly on such representations. In Statistics and Machine Learning, this problem is solved by **Hidden Markov Models (HMMs)** [33].

HMMs are a widely used model in Machine Learning, powering sequence analysis applications like speech recognition [34], handwriting recognition, and bioinformatics. They are also essential components of communication systems as well as of intelligent agents trained by reinforcement learning methods. In this article, **we describe a reaction network scheme which implements the Baum-Welch algorithm**. The Baum-Welch algorithm is an iterative algorithm for learning HMM parameters. Reaction networks that can do such statistical analysis on sequences are likely to be an essential component of molecular communication systems, enabling cooperative behavior among a population of artificial cells. Our main contributions are:

1. In Sect. 2, we describe what the reader needs to know about HMMs and the Baum-Welch algorithm to be able to follow the subsequent constructions. No prerequisites are assumed beyond familiarity with matrices and probability distributions.
2. In Sect. 3, we describe a novel reaction network scheme to learn parameters for an HMM.

3. We prove in Theorem 1 that every fixed point of the Baum-Welch algorithm is also a fixed point of the continuous dynamics of this reaction network scheme.
4. In Theorem 2, we prove that every positive fixed point of the dynamics of our reaction network scheme is a fixed point of the Baum-Welch algorithm.
5. In Theorems 3 and 4, we prove that subsets of our reaction network scheme which correspond to the Expectation step and the Maximization step of the Baum-Welch algorithm both separately converge exponentially fast.
6. In Example 1, we simulate our reaction network scheme on an input sequence and show that the network dynamics is successfully able to learn the same parameters as the Baum-Welch algorithm.
7. If observations are not taken for long enough, or if the target HMM generating the observations has zero values for some parameters, the Baum-Welch algorithm may have no positive fixed points. In this situation, the conditions for Theorem 2 to hold are not met. We show in Example 2 that the requirements of Theorem 2 are not superfluous. When the Baum-Welch algorithm has no positive fixed points, but has a boundary fixed point, our reaction network scheme can get stuck at a boundary fixed point that is not a Baum-Welch fixed point.

2   Hidden Markov Models and the Baum Welch Algorithm

Fix two finite sets P and Q. A **stochastic map** is a $|P| \times |Q|$ matrix $A = (a_{pq})_{|P| \times |Q|}$ such that $a_{pq} \geq 0$ for all $p \in P$ and $q \in Q$, and $\sum_{q \in Q} a_{pq} = 1$ for all $p \in P$. Intuitively, stochastic maps represent conditional probability distributions.

An **HMM** $(H, V, \theta, \psi, \pi)$ consists of finite sets H (for 'hidden') and V (for 'visible'), a stochastic map θ from H to H called the **transition matrix**, a stochastic map ψ from H to V called the **emission matrix**, and an **initial probability distribution** $\pi = (\pi_h)_{h \in H}$ on H, i.e., $\pi_h \geq 0$ for all $h \in H$ and $\sum_{h \in H} \pi_h = 1$. See Fig. 1a for an example.

Suppose a length L sequence $(v_1, v_2, \ldots, v_L) \in V^L$ of visible states is observed due to a hidden sequence $(x_1, x_2, \ldots, x_L) \in H^L$. The following questions related to an HMM are commonly studied:

1. **Likelihood:** For fixed θ, ψ, compute the likelihood $\Pr(v_1, v_2, \ldots, v_L \mid \theta, \psi)$. This problem is solved by the **forward-backward algorithm**.
2. **Learning:** Estimate the parameters $\hat{\theta}, \hat{\psi}$ which maximize the likelihood of the observed sequence $(v_1, v_2, \ldots, v_L) \in V^L$. This problem is solved by the **Baum-Welch algorithm** which is an Expectation-Maximization (EM) algorithm. It uses the forward-backward algorithm as a subroutine to compute the E step of EM.
3. **Decoding:** For fixed θ, ψ, find the sequence $(\hat{h}_1, \hat{h}_2, \ldots, \hat{h}_l, \ldots, \hat{h}_L) \in H^L$ that has the highest probability of producing the given observed sequence $(v_1, v_2, \ldots, v_L)$. This problem is solved by the **Viterbi algorithm**.

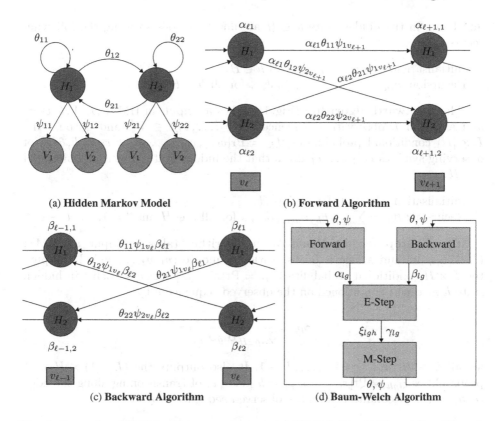

(a) Hidden Markov Model

(b) Forward Algorithm

(c) Backward Algorithm

(d) Baum-Welch Algorithm

Fig. 1. Learning HMMs from sequences. (a) HMM: The hidden states H_1 and H_2 are not directly observable. Instead what are observed are elements V_1, V_2 from the set $V = \{V_1, V_2\}$ of "visible states." The parameters $\theta_{11}, \theta_{12}, \theta_{21}, \theta_{22}$ denote the probability of transitions between the hidden states. The probability of observing states V_1, V_2 depends on the parameters $\psi_{11}, \psi_{12}, \psi_{21}, \psi_{22}$ as indicated in the figure. **(b)** The **forward algorithm** computes the position $l + 1$ likelihood $\alpha_{l+1,1} = \alpha_{l1}\theta_{11}\psi_{1v_{l+1}} + \alpha_{l2}\theta_{21}\psi_{1v_{l+1}}$ by forward propagating the position l likelihoods α_{l1} and α_{l2}. Here $v_l, v_{l+1} \in V$ are the observed emissions at position l and $l + 1$. **(c)** The **backward algorithm** computes the position $l - 1$ conditional probability $\beta_{l-1,1} = \theta_{11}\psi_{1v_l}\beta_{l1} + \theta_{12}\psi_{2v_l}\beta_{l2}$ by propagating the position l conditional probabilities β_{l1} and β_{l2} backwards. **(d)** The **Baum-Welch Algorithm** is a fixed point Expectation-Maximization computation. The E step calls the forward and backward algorithm as subroutines and, conditioned on the entire observed sequence $(v_1, v_2, \ldots, v_L) \in V^L$, computes the probabilities γ_{lg} of being in states $g \in H$ at position l and the probabilities ξ_{lgh} of taking the transitions $gh \in H^2$ at position l. The M step updates the parameters θ and ψ to maximize their likelihood given the observed sequence.

The **forward algorithm** (Fig. 1b) takes as input an HMM $(H, V, \theta, \psi, \pi)$ and a length L observation sequence $(v_1, v_2, \ldots, v_L) \in V^L$ and outputs the $L \times |H|$ likelihoods $\alpha_{lh} = \Pr[v_1, v_2, \ldots, v_l, x_l = h \mid \theta, \psi]$ of observing symbols $v_1, \ldots, v_l$

and being in the hidden state $h \in H$ at time l. It does so using the following recursion.

- Initialisation: $\alpha_{1h} = \pi_h \psi_{hv_1}$ for all $h \in H$,
- Recursion: $\alpha_{lh} = \sum_{g \in H} \alpha_{l-1,g} \theta_{gh} \psi_{hv_l}$ for all $h \in H$ and $l = 2, \ldots, L$.

The **backward algorithm** (Fig. 1c) takes as input an HMM $(H, V, \theta, \psi, \pi)$ and a length L observation sequence $(v_1, v_2, \ldots, v_L) \in V^L$ and outputs the $L \times |H|$ conditional probabilities $\beta_{lh} = \Pr[v_{l+1}, v_{l+2}, \ldots, v_L \mid x_l = h, \theta, \psi]$ of observing symbols $v_{l+1}, \ldots, v_L$ given that the hidden state x_l at time l has label $h \in H$.

- Initialisation: $\beta_{Lh} = 1$, for all $h \in H$,
- Recursion: $\beta_{lh} = \sum_{g \in H} \theta_{hg} \psi_{gv_{l+1}} \beta_{l+1,g}$ for all $h \in H$ and $l = 1, \ldots, L - 1$.

The **E step** for the Baum-Welch algorithm takes as input an HMM $(H, V, \theta, \psi, \pi)$ and a length L observation sequence $(v_1, v_2, \ldots, v_L)$. It outputs the $L \times H$ conditional probabilities $\gamma_{lh} = \Pr[x_l = h \mid \theta, \psi, v]$ of being in hidden state h at time l conditioned on the observed sequence v by:

$$\gamma_{lh} = \frac{\alpha_{lh} \beta_{lh}}{\sum_{g \in H} \alpha_{lg} \beta_{lg}}$$

for all $h \in H$ and $l = 1, 2, \ldots, L - 1$. It also outputs the $(L - 1) \times H \times H$ probabilities $\xi_{lgh} = \Pr[x_l = g, x_{l+1} = h \mid \theta, \psi, v]$ of transitioning along the edge (g, h) at time l conditioned on the observed sequence v by:

$$\xi_{lgh} = \frac{\alpha_{lg} \theta_{gh} \psi_{hv_{l+1}} \beta_{l+1,h}}{\sum_{f \in H} \alpha_{lf} \beta_{lf}}$$

for all $g, h \in H$ and $l = 1, \ldots, L - 1$.

Remark 1. Note that the E-step uses the forward and backward algorithms as subroutines to first compute the α and β values. Further note that we don't need the forward and backward algorithms to return the actual values α_{lh} and β_{lh}. To be precise, let $\alpha_l = (\alpha_{lh})_{h \in H} \in \mathbb{R}^H$ denote the vector of forward likelihoods at time l. Then for the E step to work, we only need the direction of α_l and not the magnitude. This is because the numerator and denominator in the updates are both linear in α_{lh}, and the magnitude cancels out. Similarly, if $\beta_l = (\beta_{lh})_{h \in H} \in \mathbb{R}^H$ denotes the vector of backward likelihoods at time l then the E step only cares about the direction of β_l and not the magnitude. This scale symmetry is a useful property for numerical solvers. We will also make use of this freedom when we implement a lax forward-backward algorithm using reaction networks in the next section.

The **M step** of the Baum-Welch algorithm takes as input the values γ and ξ that are output by the E step as reconstruction of the dynamics on hidden states, and outputs new Maximum Likelihood estimates of the parameters θ, ψ

that best explain these values. The update rule turns out to be very simple. For all $g, h \in H$ and $w \in V$:

$$\theta_{gh} \leftarrow \frac{\sum_{l=1}^{L-1} \xi_{lgh}}{\sum_{l=1}^{L-1} \sum_{f \in H} \xi_{lgf}}, \qquad \psi_{hw} \leftarrow \frac{\sum_{l=1}^{L} \gamma_{lh} \delta_{w,v_l}}{\sum_{l=1}^{L} \gamma_{lh}}$$

where $\delta_{w,v_l} = \begin{cases} 1 \text{ if } w = v_l \\ 0 \text{ otherwise} \end{cases}$ is the Dirac delta function.

Remark 2. Like in Remark 1, note that the M step does not require its inputs to be the actual values γ and ξ. There is a scaling symmetry so that we only need the directions of the vectors $\gamma(h) = (\gamma_{lh})_{l=1,2,\dots,L} \in \mathbb{R}^L$ for all $h \in H$ and $\xi(g) = (\xi_{lgh})_{1 \le l \le L-1, h \in H} \in \mathbb{R}^{(L-1) \times H}$ for all $g \in H$. This gives us the freedom to implement a lax E projection without affecting the M projection, and we will exploit this freedom when designing our reaction network.

The **Baum-Welch algorithm** (Fig. 1d) is a fixed point EM computation that alternately runs the E step and the M step till the updates become small enough. It is guaranteed to converge to a fixed point $(\hat{\theta}, \hat{\psi})$. However, the fixed point need not always be a global optimum.

3   Chemical Baum-Welch Algorithm

3.1   Reaction Networks

Following [25], we recall some concepts from reaction network theory [24, 35–39].

Fix a finite set S of species. An S-reaction, or simply a **reaction** when S is understood from context, is a formal chemical equation

$$\sum_{X \in S} y_X X \to \sum_{X \in S} y'_X X$$

where the numbers $y_X, y'_X \in \mathbb{Z}_{\ge 0}$ are the **stoichiometric coefficients** of species X on the **reactant** side and **product** side respectively. A **reaction network** is a pair (S, R) where R is a finite set of S-reactions. A **reaction system** is a triple (S, R, k) where (S, R) is a reaction network and $k : R \to \mathbb{R}_{>0}$ is called the **rate function**.

As is common when specifying reaction networks, we will find it convenient to explicitly specify only a set of chemical equations, leaving the set of species to be inferred by the reader.

Fix a reaction system (S, R, k). **Deterministic Mass Action Kinetics** describes a system of ordinary differential equations on the concentration variables $\{x_i(t) \mid i \in S\}$ according to:

$$\dot{x}_i(t) = \sum_{a \to b \in R} k_{a \to b}(b_i - a_i) \prod_{j \in S} x_j(t)^{a_j}$$

3.2   Baum-Welch Reaction Network

In this section we will describe a reaction network for each part of the Baum-Welch algorithm.

Fix an HMM $\mathcal{M} = (H, V, \theta, \psi, \pi)$. Pick an arbitrary hidden state $h^* \in H$ and an arbitrary visible state $v^* \in V$. Picking these states $h^* \in H$ and $v^* \in V$ is merely an artifice to break symmetry akin to selecting leaders, and our results hold independent of these choices. Also fix a length $L \in \mathbb{Z}_{>0}$ for the observed sequence.

We first work out in full detail how the forward algorithm of the Baum-Welch algorithm may be translated into chemical reactions. Suppose a length L sequence $(v_1, v_2, \ldots, v_L) \in V^L$ of visible states is observed. Then recall that the forward algorithm uses the following recursion:

- Initialisation: $\alpha_{1h} = \pi_h \psi_{hv_1}$ for all $h \in H$,
- Recursion: $\alpha_{lh} = \sum_{g \in H} \alpha_{l-1,g} \theta_{gh} \psi_{hv_l}$ for all $h \in H$ and $l = 2, \ldots, L$.

Notice this implies

- $\alpha_{1h} \times \pi_{h^*} \psi_{h^* v_1} = \alpha_{1h^*} \times \pi_h \psi_{hv_1}$ for all $h \in H \setminus \{h^*\}$,
- $\alpha_{lh} \times \left(\sum_{g \in H} \alpha_{l-1,g} \theta_{gh^*} \psi_{h^* v_l} \right) = \alpha_{lh^*} \times \left(\sum_{g \in H} \alpha_{l-1,g} \theta_{gh} \psi_{hv_l} \right)$ for all $h \in H \setminus \{h^*\}$ and $l = 2, \ldots, L$.

This prompts the use of the following reactions for the initialization step:

$$\alpha_{1h} + \pi_{h^*} + \psi_{h^* v_1} \rightarrow \alpha_{1h^*} + \pi_{h^*} + \psi_{h^* v_1}$$
$$\alpha_{1h^*} + \pi_h + \psi_{hv_1} \rightarrow \alpha_{1h} + \pi_h + \psi_{hv_1}$$

for all $h \in H \setminus \{h^*\}$ and $w \in V$. By design $\alpha_{1h} \times \pi_{h^*} \psi_{h^* v_1} = \alpha_{1h^*} \times \pi_h \psi_{hv_1}$ is the balance equation for the pair of reactions corresponding to each $h \in H \setminus \{h^*\}$.

Similarly for the recursion step, we use the following reactions:

$$\alpha_{lh} + \alpha_{l-1,g} + \theta_{gh^*} + \psi_{h^* v_l} \rightarrow \alpha_{lh^*} + \alpha_{l-1,g} + \theta_{gh^*} + \psi_{h^* v_l}$$
$$\alpha_{lh^*} + \alpha_{l-1,g} + \theta_{gh} + \psi_{hv_l} \rightarrow \alpha_{lh} + \alpha_{l-1,g} + \theta_{gh} + \psi_{hv_l}$$

for all $g \in H$, $h \in H \setminus \{h^*\}, l = 2, \ldots, L$ and $w \in V$. Again by design

$$\alpha_{lh} \times \left(\sum_{g \in H} \alpha_{l-1,g} \theta_{gh^*} \psi_{h^* v_l} \right) = \alpha_{lh^*} \times \left(\sum_{g \in H} \alpha_{l-1,g} \theta_{gh} \psi_{hv_l} \right)$$

is the balance equation for the set of reactions corresponding to each $(h, l) \in H \setminus \{h^*\} \times \{2, \ldots, L\}$.

The above reactions depend on the observed sequence $(v_1, v_2, \ldots, v_L) \in V^L$ of visible state. This is a problem because one would have to design different reaction networks for different observed sequences. To solve this problem we introduce the species E_{lw} with $l = 1, \ldots, L$ and $w \in V$. Now with the E_{lw} species, we use the following reactions for the forward algorithm:

$$\alpha_{1h} + \pi_{h^*} + \psi_{h^*w} + E_{1w} \rightarrow \alpha_{1h^*} + \pi_{h^*} + \psi_{h^*w} + E_{1w}$$

$$\alpha_{1h^*} + \pi_h + \psi_{hw} + E_{1w} \rightarrow \alpha_{1h} + \pi_h + \psi_{hw} + E_{1w}$$

for all $h \in H \setminus \{h^*\}$ and $w \in V$.

$$\alpha_{lh} + \alpha_{l-1,g} + \theta_{gh^*} + \psi_{h^*w} + E_{1w} \rightarrow \alpha_{lh^*} + \alpha_{l-1,g} + \theta_{gh^*} + \psi_{h^*w} + E_{1w}$$

$$\alpha_{lh^*} + \alpha_{l-1,g} + \theta_{gh} + \psi_{hw} + E_{1w} \rightarrow \alpha_{lh} + \alpha_{l-1,g} + \theta_{gh} + \psi_{hw} + E_{1w}$$

for all $g \in H$, $h \in H \setminus \{h^*\}, l = 2, \ldots, L$ and $w \in V$. The E_{lw} species are to be initialized such that $E_{lw} = 1$ iff $w = v_l$ and $E_{lw} = 0$ otherwise. So different observed sequences can now be processed by the same reaction network, by appropriately initializing the species E_{lw}.

The other parts of the Baum-Welch algorithm may be translated into chemical reactions using a similar logic. We call the resulting reaction network the **Baum-Welch Reaction Network** $BW(\mathcal{M}, h^*, v^*, L)$. It consists of four subnetworks corresponding to the four parts of the Baum-Welch algorithm, as shown in Table 1.

The Baum-Welch reaction network described above has a special structure. Every reaction is a monomolecular transformation catalyzed by a set of species. The reverse transformation is also present, catalyzed by a different set of species to give the network a "futile cycle" [40] structure. In addition, each connected component in the undirected graph representation of the network has a topology with all transformations happening to and from a central species. This prompts the following definitions.

Definition 1 (Flowers, petals, gardens). *A **graph** is a triple (Nodes, Edges, π) where Nodes and Edges are finite sets and π is a map from Edges to unordered pairs of elements from Nodes. A **flower** is a graph with a special node n^* such that for every edge $e \in$ Edges we have $n^* \in \pi(e)$. A **garden** is a graph which is a union of disjoint flowers. A **petal** is a set of all edges e which have the same $\pi(e)$, i.e. they are incident between the same pair of nodes.*

Figure 2 shows how the Baum-Welch reaction network can be represented as a garden graph, in which species are nodes and reactions are edges.

A collection of specific rates is **permissible** if all reactions in a petal have the same rate. However, different petals may have different rates. We will denote the specific rate for a petal by superscripting the type of the species and subscripting its indices. For example, the specific rate for reactions in the petal for α_{lh} would be denoted as k_{lh}^α. The notation for the remaining rate constants can be read from Fig. 2.

Remark 3. The "flower" topology we have employed for the Baum-Welch reaction network is only one among several possibilities. The important thing is to achieve connectivity between different nodes that ought to be connected, ensuring that the ratio of the concentrations of the species denoted by adjacent nodes

Table 1. Baum-Welch Reaction Network: The steps and reactions are for all $g, h \in H, w \in V$ and $l = 1, \ldots, L - 1$. Notice there are some null reactions of the form $a_i X_i \rightarrow a_i X_i$, which have no effect on the dynamics, and so can be ignored.

Baum-Welch Algorithm	Baum-Welch Reaction Network
$\alpha_{1h} = \pi_h \psi_{hv_1}$	$\alpha_{1h} + \pi_{h^*} + \psi_{h^*w} + E_{1w} \longrightarrow \alpha_{1h^*} + \pi_{h^*} + \psi_{h^*w} + E_{1w}$ $\alpha_{1h^*} + \pi_h + \psi_{hw} + E_{1w} \longrightarrow \alpha_{1h} + \pi_h + \psi_{hw} + E_{1w}$
$\alpha_{l+1,h} = \sum_{g \in H} \alpha_{lg} \theta_{gh} \psi_{hv_{l+1}}$	$\alpha_{l+1,h} + \alpha_{lg} + \theta_{gh^*} + \psi_{h^*w} + E_{l+1,w} \longrightarrow$ $\alpha_{l+1,h^*} + \alpha_{lg} + \theta_{gh^*} + \psi_{h^*w} + E_{l+1,w}$ $\alpha_{l+1,h^*} + \alpha_{lg} + \theta_{gh} + \psi_{hw} + E_{l+1,w} \longrightarrow$ $\alpha_{l+1,h} + \alpha_{lg} + \theta_{gh} + \psi_{hw} + E_{l+1,w}$
$\beta_{Lh} = 1$ $\beta_{lh} = \sum_{g \in H} \theta_{hg} \psi_{gv_{l+1}} \beta_{l+1,g}$	$\beta_{lh} + \beta_{l+1,g} + \theta_{h^*g} + \psi_{gw} + E_{l+1,w} \longrightarrow$ $\beta_{lh^*} + \beta_{l+1,g} + \theta_{h^*g} + \psi_{gw} + E_{l+1,w}$ $\beta_{lh^*} + \beta_{l+1,g} + \theta_{hg} + \psi_{gw} + E_{l+1,w} \longrightarrow$ $\beta_{lh} + \beta_{l+1,g} + \theta_{hg} + \psi_{gw} + E_{l+1,w}$
$\gamma_{lh} = \dfrac{\alpha_{lh} \beta_{lh}}{\sum_{g \in H} \alpha_{lg} \beta_{lg}}$	$\gamma_{lh} + \alpha_{lh^*} + \beta_{lh^*} \longrightarrow \gamma_{lh^*} + \alpha_{lh^*} + \beta_{lh^*}$ $\gamma_{lh^*} + \alpha_{lh} + \beta_{lh} \longrightarrow \gamma_{lh} + \alpha_{lh} + \beta_{lh}$
$\xi_{lgh} = \dfrac{\alpha_{lg} \theta_{gh} \psi_{hv_{l+1}} \beta_{l+1,h}}{\sum_{f \in H} \alpha_{lf} \beta_{lf}}$	$\xi_{lgh} + \alpha_{lh^*} + \theta_{h^*h^*} + \beta_{l+1,h^*} + \psi_{h^*w} + E_{l+1,w} \longrightarrow$ $\xi_{lh^*h^*} + \alpha_{lh^*} + \theta_{h^*h^*} + \beta_{l+1,h^*} + \psi_{h^*w} + E_{l+1,w}$ $\xi_{lh^*h^*} + \alpha_{lg} + \theta_{gh} + \beta_{l+1,g} + \psi_{hw} + E_{l+1,w} \longrightarrow$ $\xi_{lgh} + \alpha_{lg} + \theta_{gh} + \beta_{l+1,g} + \psi_{hw} + E_{l+1,w}$
$\theta_{gh} \leftarrow \dfrac{\sum_{l=1}^{L-1} \xi_{lgh}}{\sum_{l=1}^{L-1} \sum_{f \in H} \xi_{lgf}}$	$\theta_{gh} + \xi_{lgh^*} \longrightarrow \theta_{gh^*} + \xi_{lgh^*}$ $\theta_{gh^*} + \xi_{lgh} \longrightarrow \theta_{gh} + \xi_{lgh}$
$\psi_{hw} \leftarrow \dfrac{\sum_{l=1}^{L} \gamma_{lh} \delta_{w,v_l}}{\sum_{l=1}^{L} \gamma_{lh}}$	$\psi_{hw} + \gamma_{lh} + E_{lv^*} \longrightarrow \psi_{hv^*} + \gamma_{lh} + E_{lv^*}$ $\psi_{hv^*} + \gamma_{lh} + E_{lw} \longrightarrow \psi_{hw} + \gamma_{lh} + E_{lw}$

takes the value as prescribed by the Baum-Welch algorithm. Therefore, many other connection topologies can be imagined, for example a ring topology where each node is connected to two other nodes while maintaining the same connected

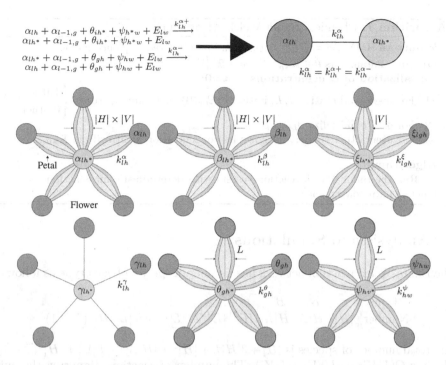

Fig. 2. The Baum-Welch Reaction Network represented as an undirected graph. (a) Each reaction is a unimolecular transformation driven by catalysts. The nodes represents the species undergoing transformation. The edge represents a pair of reactions which drive this transformation backwards and forwards in a futile cycle. Catalysts are omitted for clarity. (b) The network decomposes into a collection of disjoint **flowers**. Nodes represent species and edges represent pairs of reactions, species α, γ have L flowers each, β, ξ have $L - 1$ flowers each, and species θ and ψ have $|H|$ flowers each (not shown in figure). All reactions in the same petal have the same specific rate constant, so the dimension of the space of permissible rate constants is equal to the number of petals in the graph.

components. In fact, there are obvious disadvantages to the star topology, with a single point of failure, whereas the ring topology appears more resilient. How network topology affects basins of attraction, rates of convergence, and emergence of spurious equilibria in our algorithm is a compelling question beyond the scope of this present work.

The Baum-Welch reaction network with a permissible choice of rate constants k will define a **Baum-Welch reaction system** $(BW(\mathcal{M}, h^*, v^*, L), k)$ whose deterministic mass action kinetics equations will perform a continuous time version of the Baum-Welch algorithm. We call this the **Chemical Baum-Welch Algorithm**, and describe it in Algorithm 1.

Algorithm 1. Chemical Baum-Welch Algorithm

Input: An HMM $\mathcal{M} = (H, V, \theta, \psi, \pi)$ and an observed sequence $v \in V^L$
Output: Parameters $\hat{\theta} \in \mathbb{R}_{\geq 0}^{|H| \times |H|}, \hat{\psi} \in \mathbb{R}_{\geq 0}^{|H| \times |V|}$
Initialization of concentrations at t=0:

1. For $w \in V$ and $l = 1, \ldots, L$, initialize $E_{lw}(0)$ such that $E_{lw}(0) = \begin{cases} 1 \text{ if } w = v_l \\ 0 \text{ otherwise} \end{cases}$

2. For $g \in H$, initialize $\beta_{Lg} = \beta$

3. For every other species, initialize its concentration arbitrarily in $\mathbb{R}_{>0}$.

Algorithm:
   Run the Baum-Welch reaction system with deterministic mass action kinetics until convergence.

---

4   Analysis and Simulations

The Baum-Welch reaction network has number of species of each type as follows:

Type	π	α	β	θ	ψ	E	γ	ξ
Number	$|H|$	$|H|L$	$|H|L$	$|H|^2$	$|H||V|$	$L|V|$	$|H|L$	$|H|^2(L-1)$

The total number of species is $|H| + 3|H|L + |H|^2 + |H||V| + L|V| + |H|^2(L-1)$ which is $O(L|H|^2 + |H||V| + L|V|)$. The number of reactions (ignoring the null reactions of the form $\sum a_i X_i \to \sum a_i X_i$) in each part is:

Forward	Backward	Expectation	Maximization
$2(|H|-1)V$	$(2|H|(|H|-1)|V|)$	$2L(|H|-1)$	$2|H|(|H|-1)(L-1)$
$+2|H|(|H|-1)(L-1)|V|$	$(L-1)$	$+2(L-1)(|H|^2-1)$	$+2|H|L(|V|-1)$

so that the total number of reactions is $O(|H|^2L|V|)$.

   The first theorem shows that the Chemical Baum-Welch Algorithm recovers all of the Baum-Welch equilibria.

Theorem 1. *Every fixed point of the Baum-Welch algorithm for an HMM $\mathcal{M} = (H, V, \theta, \psi, \pi)$ is a fixed point for the corresponding Chemical Baum-Welch Algorithm with permissible rates k.*

See Appendix A.1 for proof.
   We say a vector of real numbers is **positive** if all its components are strictly greater than 0. The next theorem shows that positive equilibria of the Chemical Baum-Welch Algorithm are also fixed points of the Baum-Welch algorithm.

Theorem 2. *Every **positive** fixed point for the Chemical Baum-Welch Algorithm on a Baum-Welch Reaction system $(BW(\mathcal{M}, h^*, v^*, L), k)$ with permissible rate k is a fixed point for the Baum-Welch algorithm for the HMM $\mathcal{M} = (H, V, \theta, \psi, \pi)$.*

See Appendix A.1 for proof.

The Baum-Welch algorithm is an iterative algorithm. The E step and the M step are run iteratively. In contrast, the Chemical Baum-Welch algorithm is a generalized EM algorithm [41] where all the reactions are run at the same time in a single-pot reaction.

The next theorem shows that the E step consisting of the forward network, the backward network, and the E network, converges exponentially fast to the correct equilibrium if the θ and ψ species are held fixed at a *positive* point.

Theorem 3. *For the Baum-Welch Reaction System* $(BW(\mathcal{M}, h^*, v^*, L), k)$ *with permissible rates k, if the concentrations of θ and ψ species are held fixed at a positive point then the Forward, Backward and Expection step reaction systems on α, β, γ and ψ species converge to equilibrium exponentially fast.*

See Appendix A.2 for proof.

The next theorem shows that if the $\alpha, \beta, \gamma, \xi$ species are held fixed at a *positive* point, then the M step consisting of reactions modifying the θ and ψ species converges exponentially fast to the correct equilibrium.

Theorem 4. *For the Baum-Welch Reaction System* $(BW(\mathcal{M}, h^*, v^*, L), k)$ *with permissible rates k, if the concentrations of α, β, γ and ξ species are held fixed at a positive point then the Maximization step reaction system on θ and ψ converges to equilibrium exponentially fast.*

See Appendix A.2 for proof.

The following examples demonstrate the behavior of the Chemical Baum-Welch Algorithm.

Example 1. Consider an HMM $(H, V, \theta, \psi, \pi)$ with two hidden states $H = \{H_1, H_2\}$ and two emitted symbols $V = \{V_1, V_2\}$ where the starting probability is $\pi = (0.6, 0.4)$, initial transition probability is $\theta = \begin{bmatrix} 0.6 & 0.4 \\ 0.3 & 0.7 \end{bmatrix}$, and initial emission probability is $\psi = \begin{bmatrix} 0.5 & 0.5 \\ 0.5 & 0.5 \end{bmatrix}$. Suppose we wish to learn (θ, ψ) for the following observed sequence: $(V_1, V_1, V_1, V_2, V_1, V_1, V_2, V_2, V_2, V_1, V_2, V_2, V_1, V_1, V_1, V_2, V_2, V_2, V_1, V_2, V_1, V_1, V_2, V_2)$. We initialize the corresponding reaction system by setting species $E_{l,v_l} = 1$ and $E_{l,w} = 0$ for $w \neq v_l$, and run the dynamics according to deterministic mass-action kinetics. Initial conditions of all species that are not mentioned are chosen to be nonzero, but otherwise at random.

For our numerical solution, we observe that the reaction network equilibrium point coincides with the Baum-Welch steady state $\hat{\theta} = \begin{bmatrix} 0.5071 & 0.4928 \\ 0.0000 & 1.0000 \end{bmatrix}$ and $\hat{\psi} = \begin{bmatrix} 1.0000 & 0.0000 \\ 0.4854 & 0.5145 \end{bmatrix}$ (See Fig. 3).

The next example shows that the Chemical Baum-Welch algorithm can sometimes get stuck at points that are not equilibria for the Baum-Welch algorithm.

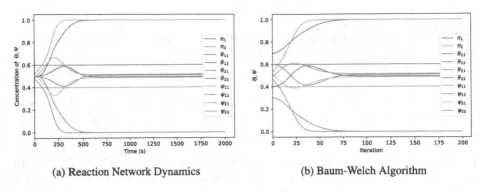

(a) Reaction Network Dynamics   (b) Baum-Welch Algorithm

Fig. 3. Simulation of Example 1: Both simulations are started at exactly the same initial vector. This may not be apparent in the figure because concentration of some species change rapidly at start in the reaction network.

This is a problem especially for very short sequences, and is probably happening because in such settings, the best model sets many parameters to zero. When many species concentrations are set to 0, many reactions get turned off, and the network gets stuck away from the desired equilibrium. We believe this will not happen if the HMM generating the observed sequence has nonzero parameters, and the observed sequence is long enough.

Example 2. Consider an HMM $(H, V, \theta, \psi, \pi)$ with two hidden states $H = \{H_1, H_2\}$ and two emitted symbols $V = \{V_1, V_2\}$, initial distribution $\pi = (0.6, 0.4)$ is fixed, initial transition probability $\theta = \begin{bmatrix} 0.6 & 0.4 \\ 0.3 & 0.7 \end{bmatrix}$, and initial emission probability $\psi = \begin{bmatrix} 0.5 & 0.5 \\ 0.5 & 0.5 \end{bmatrix}$. Suppose we wish to learn (θ, ψ) for the sequence $(V_1, V_2, V_1, V_2, V_1)$. We again simulate the corresponding reaction network and also perform the Baum-Welch algorithm for comparison.

Figure 4 shows that the reaction network equilibrium point does not coincide with the Baum-Welch steady state. Both converge to $\hat{\psi} = \begin{bmatrix} 1.0000 & 0.0000 \\ 0.0000 & 1.0000 \end{bmatrix}$. However, the reaction network converges to $\hat{\theta} = \begin{bmatrix} 0.3489 & 0.6510 \\ 1.0000 & 0.0000 \end{bmatrix}$ whereas the Baum-Welch algorithm converges to $\hat{\theta} = \begin{bmatrix} 0.0000 & 1.0000 \\ 1.0000 & 0.0000 \end{bmatrix}$ which happens to be the true maximum likelihood point. Note that this does not contradict Theorem 2 because the fixed point of this system is a boundary point and not a positive fixed point.

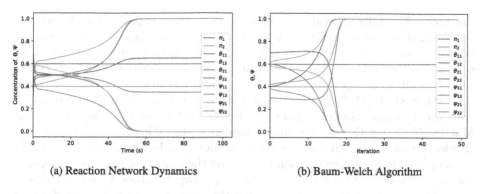

(a) Reaction Network Dynamics   (b) Baum-Welch Algorithm

Fig. 4. Simulation of Example 2: Both simulations are started at exactly the same initial vector. This may not be apparent in the figure because concentration of some species change rapidly at start in the reaction network.

5 Related Work

In previous work, our group has shown that reaction networks can perform the Expectation Maximization (EM) algorithm for partially observed log linear statistical models [25,42]. That algorithm also applies "out of the box" to learning HMM parameters. The problem with that algorithm is that the size of the reaction network would become exponentially large in the length of the sequence, so that even examples like Example 1 with an observation sequence of length 25 would become impractical. In contrast, the scheme we have presented in this paper requires only a linear growth with sequence length. We have obtained the savings by exploiting the graphical structure of HMMs. This allows us to compute the likelihoods α in a "dynamic programming" manner, instead of having to explicitly represent each path as a separate species.

Napp and Adams [20] have shown how to compute marginals on graphical models with reaction networks. They exploit graphical structure by mimicking belief propagation. Hidden Markov Models can be viewed as a special type of graphical model where there are $2L$ random variables $X_1, X_2, \ldots, X_L, Y_1, Y_2, \ldots, Y_L$ with the X random variables taking values in H and the Y random variables in V. The X random variables form a Markov chain $X_1 - X_2 - \ldots - X_L$. In addition, there are L edges from X_l to Y_l for $l = 1$ to L denoting observations. Specialized to HMMs, the scheme of Napp and Adams would compute the equivalent of steady state values of the γ species, performing a version of the E step. They are able to show that true marginals are fixed points of their scheme, which is similar to our Theorem 1. Thus their work may be viewed as the first example of a reaction network scheme that exploits graphical structure to compute E projections. Our E step goes further by proving correctness as well as exponential convergence. Their work also raises the challenge of extending our scheme to all graphical models.

Poole et al. [43] have described Chemical Boltzmann Machines, which are reaction network schemes whose dynamics reconstructs inference in Boltzmann

Machines. This inference can be viewed as a version of E projection. No scheme for learning is presented. The exact schemes presented there are exponentially large. The more realistically sized schemes are presented there without proof. In comparison, our schemes are polynomially sized, provably correct if the equilibrium is positive, and perform both inference and learning for HMMs.

Zechner et al. [11] have shown that Kalman filters can be implemented with reaction networks. Kalman filters can be thought of as a version of Hidden Markov Models with a continuum of hidden states [44]. It would be instructive to compare their scheme with ours, and note similarities and differences. In passing from position l to position $l + 1$ along the sequence, our scheme repeats the same reaction network that updates α_{l+1} using α_l values. It is worth examining if this can be done "in place" so that the same species can be reused, and a reaction network can be described that is not tied to the length L of the sequence to be observed.

Recently Cherry et al. [10] have given a brilliant experimental demonstration of learning with DNA molecules. They have empirically demonstrated a DNA molecular system that can classify 9 types of handwritten digits from the MNIST database. Their approach is based on the notion of "winner-takes-all" circuits due to Maass [45] which was originally a proposal for how neural networks in the brain work. Winner-take-all might also be capable of approximating HMM learning, at least in theory [46], and it is worth understanding precisely how such schemes relate to the kind of scheme we have described here. It is conceivable that our scheme could be converted to winner-take-all by getting different species in the same flower to give negative feedback to each other. This might well lead to sampling the most likely path, performing a decoding task similar to the Viterbi algorithm.

6   Discussion

We have described a one-pot one-shot reaction network implementation of the Baum-Welch algorithm. Firstly, this involves proposing a reaction system whose positive fixed points correspond to equilibria of the Baum-Welch Algorithm. Secondly, this involves establishing the conditions under which convergence of solutions to the equilibria can be accomplished. Further, from a practical perspective, it is essential to obtain an implementation that does not rely on repeated iteration of a reaction scheme but only requires to be run once.

As we observe in Remark 3, there is a whole class of reaction networks that implements the Baum-Welch algorithm. We have proposed one such network and are aware that there are other networks, potentially with more efficient dynamical and analytical properties than the one proposed here. Finding and characterizing efficient reaction networks that can do complicated statistical tasks will likely be of future concern.

We have only discussed deterministic dynamical properties of the network. However, in realistic biological contexts one might imagine that the network is implemented by relatively few molecules such that stochastic effects are significant. Consequently, the study of the Baum-Welch reaction system under stochastic mass-action kinetics is likely to be of interest.

Lastly, we have mentioned the Viterbi algorithm, but have made no attempt to describe how the maximum likelihood sequence can be recovered from our reaction network. This decoding step is likely to be of as much interest for molecular communication systems and molecular multi-agent systems as it is in more traditional domains of communications and multi-agent reinforcement learning. Because of the inherent stochasticity of reaction networks, there might even be opportunities for list decoding by sampling different paths through the hidden states that have high probability conditioned on the observations. This might give an artificial cell the ability to "imagine" different possible realities, and act assuming one of them to be the case, leading to an intrinsic stochasticity and unpredictability to the behavior.

A   Appendix

A.1   Comparing Points of Equilibria

We will now prove Theorem 1. For the sake of convenience, we first recall the statement.

Theorem 1. *Every fixed point of the Baum-Welch algorithm for an HMM* $\mathcal{M} = (H, V, \theta, \psi, \pi)$ *is a fixed point for the corresponding Chemical Baum-Welch Algorithm with permissible rates* k.

Proof. Consider a point $\Phi = (\alpha', \beta', \gamma', \xi', \theta', \psi')$ with $\alpha', \beta', \gamma' \in \mathbb{R}_{\geq 0}^{L \times H}$, $\xi' \in \mathbb{R}_{\geq 0}^{(L-1) \times H \times H}$, $\theta' \in \mathbb{R}_{\geq 0}^{H \times H}$ and $\psi' \in \mathbb{R}_{\geq 0}^{H \times V}$. If Φ is a fixed point of Baum-Welch Algorithm then it must satisfy:

- $\alpha'_{1h} = \pi_h \psi'_{hv_1}$ for all $h \in H$. Then for the chemical Baum-Welch Algorithm we have

$$\dot{\alpha}_{1h}\big|_{\Phi} = k_{1h}^{\alpha} \left(\alpha'_{1h^*} \pi_h \psi'_{hv_1} - \alpha'_{1h} \pi_{h^*} \psi'_{h^* v_1} \right) = 0$$

for all $h \in H \setminus \{h^*\}$ and $\dot{\alpha}_{1h^*}\big|_{\Phi} = -\sum_{h \neq h^*} \dot{\alpha}_{1h}\big|_{\Phi} = 0$.
- $\alpha'_{lh} = \sum_{g \in H} \alpha'_{l-1,g} \theta'_{gh} \psi'_{hv_l}$ for all $h \in H$ and $l = 2, \ldots, L$. Then for the chemical Baum-Welch Algorithm we have

$$\dot{\alpha}_{lh}\big|_{\Phi} = k_{lh}^{\alpha} \left(\alpha'_{lh^*} \sum_{g \in H} \alpha'_{l-1,g} \theta'_{gh} \psi'_{hv_l} - \alpha'_{lh} \sum_{g \in H} \alpha'_{l-1,g} \theta'_{gh^*} \psi'_{h^* v_l} \right) = 0$$

for all $h \in H \setminus \{h^*\}$ and $l = 2, \ldots, L$ and $\dot{\alpha}_{lh^*}\big|_{\Phi} = -\sum_{h \neq h^*} \dot{\alpha}_{lh}\big|_{\Phi} = 0$.
- $\beta'_{lh} = \sum_{g \in H} \theta'_{hg} \psi'_{gv_{l+1}} \beta'_{l+1,g}$ for all $h \in H$ and $l = 1, \ldots, L-1$. Then for the chemical Baum-Welch Algorithm we have

$$\dot{\beta}_{lh}\big|_{\Phi} = k_{lh}^{\beta} \left(\beta'_{lh^*} \sum_{g \in H} \theta'_{h^* g} \psi'_{gv_{l+1}} \beta'_{l+1,g} - \beta'_{lh} \sum_{g \in H} \theta'_{hg} \psi'_{gv_{l+1}} \beta'_{l+1,g} \right) = 0$$

for all $h \in H \setminus \{h^*\}$ and $l = 1, 2, \ldots, L-1$ and $\dot{\beta}_{lh^*}\big|_{\Phi} = -\sum_{h \neq h^*} \dot{\beta}_{lh}\big|_{\Phi} = 0$.

- $\gamma'_l(h) = \frac{\alpha'_{lh}\beta'_{lh}}{\sum_{g\in H}\alpha'_{lg}\beta'_{lg}}$ for all $h \in H$ and $l = 1,2,\ldots,L-1$. Then for the chemical Baum-Welch Algorithm we have

$$\dot{\gamma}_{lh} = k^{\gamma}_{lh}\left(\gamma'_{lh^*}\alpha'_{lh}\beta'_{lh} - \gamma'_{lh}\alpha'_{lh^*}\beta'_{lh^*}\right) = 0$$

for all $h \in H \setminus \{h^*\}$ and $l = 1,2,\ldots,L-1$ and $\dot{\gamma}_{lh^*} = -\sum_{h\neq h^*}\dot{\gamma}_{lh} = 0$.

- $\xi'_l(g,h) = \frac{\alpha'_{lg}\theta'_{gh}\psi'_{hv_{l+1}}\beta'_{l+1,h}}{\sum_{f\in H}\alpha'_{lf}\beta'_{lf}}$ for all $g,h \in H$ and $l = 1,\ldots,L-1$. Then for the chemical Baum-Welch Algorithm we have

$$\dot{\xi}_{lgh}\big|_{\Phi} = k^{\xi}_{lgh}\left(\xi'_{lh^*h^*}\alpha'_{lg}\theta'_{gh}\psi'_{hv_{l+1}}\beta'_{l+1,h} - \xi'_{lgh}\alpha'_{lh^*}\theta'_{h^*h^*}\psi'_{h^*v_{l+1}}\beta'_{l+1,h^*}\right) = 0$$

for all $g,h \in H \times H \setminus \{(h^*,h^*)\}$ and $l = 1,\ldots,L-1$ and $\dot{\xi}_{lh^*h^*}\big|_{\Phi} = -\sum_{(g,h)\neq(h^*,h^*)}\dot{\xi}_{lgh}\big|_{\Phi} = 0$.

- $\theta'_{gh} = \frac{\sum_{l=1}^{L-1}\xi'_l(g,h)}{\sum_{l=1}^{L-1}\sum_{f\in H}\xi'_l(g,f)}$ for all $g,h \in H$. Then for the chemical Baum-Welch Algorithm we have

$$\dot{\theta}_{gh}\big|_{\Phi} = k^{\theta}_{gh}\left(\theta'_{gh^*}\sum_{l=1}^{L-1}\xi'_{lgh} - \theta'_{gh}\sum_{l=1}^{L-1}\xi'_l(g,h^*)\right) = 0$$

for all $g \in H$ and $h \in H \setminus \{h^*\}$ and $\dot{\theta}_{gh^*}\big|_{\Phi} = -\sum_{h\neq h^*}\dot{\theta}_{gh}\big|_{\Phi} = 0$.

- $\psi'_{hw} = \frac{\sum_{l=1}^{L}\gamma'_l(h)\delta_{w,v_l}}{\sum_{l=1}^{L}\gamma'_l(h)}$ for all $h \in H$ and $w \in V$. Then for the chemical Baum-Welch Algorithm we have

$$\dot{\theta}_{gh}\big|_{\Phi} = k^{\theta}_{gh}\left(\theta'_{gh^*}\sum_{l=1}^{L-1}\xi'_{lgh} - \theta'_{gh}\sum_{l=1}^{L-1}\xi'_l(g,h^*)\right) = 0$$

for all $h \in H$ and $w \in V \setminus \{v^*\}$ and $\dot{\psi}_{hw^*}\big|_{\Phi} = -\sum_{w\neq w^*}\dot{\psi}_{hw}\big|_{\Phi} = 0$.

So Φ is fixed point of the chemical Baum-Welch Algorithm.   □

We will now prove Theorem 2. For the sake of convenience, we first recall the statement.

Theorem 2. *Every positive fixed point for the Chemical Baum-Welch Algorithm on a Baum Welch Reaction system $(BW(\mathcal{M}, h^*, v^*, L), k)$ with permissible rate k is a fixed point for the Baum-Welch algorithm for the HMM $\mathcal{M} = (H, V, \theta, \psi, \pi)$.*

Proof. Consider a *positive* point $\Phi = (\alpha', \beta', \gamma', \xi', \theta', \psi')$ with $\alpha', \beta', \gamma' \in \mathbb{R}^{L\times H}_{>0}$, $\xi' \in \mathbb{R}^{(L-1)\times H\times H}_{>0}$, $\theta' \in \mathbb{R}^{H\times H}_{>0}$ and $\psi' \in \mathbb{R}^{H\times V}_{>0}$. If Φ is a fixed point for the Chemical Baum-Welch Algorithm then we must have:

- $\dot{\alpha}_{1h}\big|_{\Phi} = 0$ for all $h \in H$. This implies $\alpha'_{1h} \times \pi_{h^*}\psi'_{h^*v_1} = \alpha'_{1h^*} \times \pi_h\psi'_{hv_1}$ for all $h \in H \setminus h^*$. Since Φ is positive, this implies

$$\alpha'_{1h} = \left(\pi_h\psi'_{hv_1}\right)\frac{\sum_{f\in H}\alpha'_{1f}}{\sum_{f\in H}\pi_f\psi'_{fv_1}} \qquad \text{for all } h \in H$$

- $\dot{\alpha}_{lh}\big|_{\Phi} = 0$ for all $h \in H$ and $l = 2, \ldots, L$. This implies $\alpha'_{lh} \times \sum_{g \in H} \alpha'_{l-1,g}\theta'_{gh^*}\psi'_{h^*v_l} = \alpha'_{lh^*} \times \sum_{g \in H} \alpha'_{l-1,g}\theta'_{gh}\psi'_{hv_l}$ for all $h \in H \setminus \{h^*\}$ and $l = 2, \ldots, L$. Since Φ is positive, this implies

$$\alpha'_{lh} = \left(\sum_{g \in H} \alpha'_{l-1,g}\theta'_{gh}\psi'_{hv_l} \right) \frac{\sum_{f \in H} \alpha'_{lf}}{\sum_{f,g \in H} \alpha'_{l-1,g}\theta'_{gf}\psi'_{fv_l}} \quad \text{for all } h \in H \text{ and } l = 2, \ldots, L$$

- $\dot{\beta}_{lh}\big|_{\Phi} = 0$ for all $h \in H$ and $l = 1, \ldots, L$. This implies $\beta'_{lh} \times \sum_{g \in H} \theta'_{hg}\psi'_{gv_{l+1}}\beta'_{l+1,g} = \beta'_{lh^*} \times \sum_{g \in H} \theta'_{h^*g}\psi'_{gv_{l+1}}\beta'_{l+1,g}$ for all $h \in H \setminus \{h^*\}$ and $l = 1, \ldots, L - 1$. Since Φ is positive, this implies

$$\beta'_{lh} = \left(\sum_{g \in H} \theta'_{hg}\psi'_{gv_{l+1}}\beta'_{l+1,g} \right) \frac{\sum_{f \in H} \beta'_{lf}}{\sum_{f,g \in H} \theta'_{fg}\psi'_{gv_{l+1}}\beta'_{l+1,g}} \quad \text{for all } h \in H \text{ and } l = 1, \ldots, L - 1$$

- $\dot{\gamma}_{lh}\big|_{\Phi} = 0$ for all $h \in H$ and $l = 1, \ldots, L$. This implies $\gamma'_l(h) \times \alpha'_{lh^*}\beta'_{lh^*} = \gamma'_l(h^*) \times \alpha'_{lh}\beta'_{lh}$ for all $h \in H \setminus \{h^*\}$ and $l = 1, 2, \ldots, L - 1$. Since Φ is positive, this implies

$$\gamma'_{lh} = \left(\frac{\alpha'_{lh}\beta'_{lh}}{\sum_{g \in H} \alpha'_{lg}\beta'_{lg}} \right) \sum_{g \in H} \gamma'_{lg} \quad \text{for all } h \in H \text{ and } l = 1, 2, \ldots, L - 1$$

- $\dot{\xi}_{lgh}\big|_{\Phi} = 0$ for all $g, h \in H$ and $l = 1, \ldots, L - 1$. This implies $\xi'_l(g, h) \times \alpha'_{lh^*}\theta'_{h^*h^*}\psi'_{h^*v_{l+1}}\beta'_{l+1,h^*} = \xi'_l(h^*, h^*) \times \alpha'_{lg}\theta'_{gh}\psi'_{hv_{l+1}}\beta'_{l+1,h}$ for all $g, h \in H \times H \setminus \{(h^*, h^*)\}$ and $l = 1, \ldots, L - 1$. Since Φ is positive, this implies

$$\xi'_{lgh} = \left(\frac{\alpha'_{lg}\theta'_{gh}\psi'_{hv_{l+1}}\beta'_{l+1,h}}{\sum_{e,f \in H} \alpha'_{lf}\theta'_{ef}\psi'_{fv_{l+1}}\beta'_{l+1,f}} \right) \sum_{e,f \in H} \xi'_{lef} \quad \text{for all } g, h \in H \times H \text{ and } l = 1, \ldots, L - 1$$

- $\dot{\theta}_{gh}\big|_{\Phi} = 0$ for all $g, h \in H$. This implies $\theta'_{gh} \times \sum_{l=1}^{L-1} \xi'_l(g, h^*) = \theta'_{gh^*} \times \sum_{l=1}^{L-1} \xi'_l(g, h)$ for all $g \in H$ and $h \in H \setminus \{h^*\}$. Since Φ is positive, this implies

$$\theta'_{gh} = \left(\frac{\sum_{l=1}^{L-1} \xi'_{lgh}}{\sum_{f \in H} \sum_{l=1}^{L-1} \xi'_{lgf}} \right) \sum_{f \in H} \theta'_{gf} \quad \text{for all } g \in H \text{ and } h \in H$$

- $\dot{\psi}_{hw}\big|_{\Phi} = 0$ for all $h \in H$ and $w \in V$. This implies $\psi'_{hw} \times \sum_{l=1}^{L} \gamma'_l(h)\delta_{v^*,v_l} = \psi'_{hv^*} \times \sum_{l=1}^{L} \gamma'_l(h)\delta_{w,v_l}$ for all $h \in H$ and $w \in V \setminus \{v^*\}$. Since Φ is positive, $E_{lv} = \delta_{v,v_l}$ and $\sum_{v \in V} \delta_{v,v_l} = 1$ this implies

$$\psi'_{hw} = \left(\frac{\sum_{l=1}^{L} \gamma'_{lh}\delta_{w,v_l}}{\sum_{l=1}^{L} \gamma'_{lh}} \right) \sum_{v \in V} \psi'_{hv} \quad \text{for all } h \in H \text{ and } w \in V$$

Because of the relaxation we get by Remark 1, the point Φ qualifies as a fixed point of the Baum-Welch algorithm. $\qquad \square$

A.2   Rate of Convergence Analysis

In this section we will prove Theorems 3 and 4, but first we will state and prove two useful lemmas.

Lemma 1. *Let A be an arbitrary $n \times n$ matrix. Let W be an $r \times n$ matrix comprising of r linearly independent left kernel vectors of A so that $WA = 0_{r,n}$, where $0_{i,j}$ denotes a $i \times j$ matrix with all entries zero. Further suppose W is in the row reduced form, that is,*

$$W = \begin{pmatrix} W' & I_r \end{pmatrix}$$

where I_j denotes the $j \times j$ identity matrix and W' is a $r \times (n-r)$ matrix. Let A be given as

$$A = \begin{pmatrix} A_{11} & A_{12} \\ A_{21} & A_{22} \end{pmatrix},$$

where A_{11} is a $(n-r) \times (n-r)$ matrix, A_{12} is a $(n-r) \times r$ matrix, A_{21} is a $r \times (n-r)$ matrix, and A_{22} is a $r \times r$ matrix.

Then the $n-r$ eigenvalues (with multiplicity) of the matrix $A_{11} - A_{12}W'$ are the same as the eigenvalues of A except for r zero eigenvalues.

Proof. Consider the $n \times n$ invertible matrix P given by

$$P = \begin{pmatrix} I_{n-r} & 0_{n-r,r} \\ W' & I_r \end{pmatrix}, \quad P^{-1} = \begin{pmatrix} I_{n-r} & 0_{n-r,r} \\ -W' & I_r \end{pmatrix},$$

with determinant $\det(P) = \det(P^{-1}) = 1$. We have

$$\begin{aligned}
PAP^{-1} &= \begin{pmatrix} I_{n-r} & 0_{n-r,r} \\ W' & I_r \end{pmatrix} \begin{pmatrix} A_{11} & A_{12} \\ A_{21} & A_{22} \end{pmatrix} \begin{pmatrix} I_{n-r} & 0_{n-r,r} \\ -W' & I_r \end{pmatrix} \\
&= \begin{pmatrix} A_{11} & A_{12} \\ 0_{r,n-r} & 0_{r,r} \end{pmatrix} \begin{pmatrix} I_{n-r} & 0_{n-r,r} \\ -W' & I_r \end{pmatrix} \\
&= \begin{pmatrix} A_{11} - A_{12}W' & A_{12} \\ 0_{r,n-r} & 0_{r,r} \end{pmatrix}
\end{aligned}$$

This implies that the characteristic polynomial of A fulfils

$$\begin{aligned}
\det(A - \lambda I_n) &= \det(P)\det(A - \lambda I_n)\det(P^{-1}) = \det(PAP^{-1} - \lambda I_n) \\
&= \det \begin{pmatrix} A_{11} - A_{12}W' - \lambda I_{n-r} & A_{12} \\ 0_{r,n-r} & 0_{r,r} - \lambda I_r \end{pmatrix} \\
&= (-1)^r \lambda^r \det(A_{11} - A_{12}W' - \lambda I_{n-r}),
\end{aligned}$$

and the statement follows.   □

Now we revisit the observation that every reaction in the Baum-Welch reaction network is a monomolecular transformation catalyzed by a set of species. For the purposes of our analysis, each reaction can be abstracted as a monomolecular reaction with time varying rate constants. This prompts us to consider the following monomolecular reaction system with n species $X_1, \ldots, X_n$ and m reactions

$$X_{r_j} \xrightarrow{k_j(t)} X_{p_j}, \quad \text{for} \quad j = 1, \ldots, m,$$

where $r_j \neq p_j$, and $r_j, p_j \in \{1, \ldots, n\}$, and $k_j(t)$, $j = 1, \ldots, m$, are mass-action reaction rate constants, possibly depending on time. We assume $k_j(t) > 0$ for $t \geq 0$ and let $k(t) = (k_1(t), \ldots, k_m(t))$ be the vector of reaction rate constants. Furthermore, we assume there is at most one reaction j such that $(r_j, p_j) = (r, p) \in \{1, \ldots, n\}^2$ and that the reaction network is strongly connected. The later means there is a reaction path from any species X_i to any other species $X_{i'}$. (In reaction network terms it means the network is weakly reversible and deficiency zero.)

The mass action kinetics of this reaction system is given by the ODE system

$$\dot{x}_i = -x_i \sum_{j=1:\, r_j=i}^{m} k_j(t) + \sum_{j=1:\, p_j=i}^{m} k_j(t) x_{r_j}, \quad i = 1, \ldots, n.$$

Define the $n \times n$ matrix $A(t) = (a_{ii'}(t))_{i,i'=1,\ldots,n}$ by

$$a_{ii}(t) = -\sum_{j=1:\, r_j=i}^{m} k_j(t),$$

$$a_{ii'}(t) = k_j(t), \quad \text{if there is } j \in \{1, \ldots, m\} \text{ such that } (r_j, p_j) = (i', i). \tag{1}$$

Then the ODE system might be written as

$$\dot{x} = A(t)x. \tag{2}$$

Note that the column sums of $A(t)$ are zero, implying that $\sum_{i=1}^{n} x_i$ is conserved.

Lemma 2. *Assume $k(t)$ for $t \geq 0$, converges exponentially fast towards $k = (k_1, \ldots, k_m) \in R_{>0}^m$ as $t \to \infty$, that is, there exists $\gamma_1 > 0$ and $K_1 > 0$ such that*

$$\| k(t) - k \| \leq K_1 e^{-\gamma_1 t} \quad \text{for} \quad t \geq 0.$$

Let $A(t)$ be the matrix as defined in Eq. 1. And let A be the matrix obtained with k inserted for $k(t)$ in the matrix $A(t)$ that is, $A = \lim_{t \to \infty} A(t)$.

Then solutions of ODE system $\dot{x} = A(t)x$ starting at $x(0) \in \mathbb{R}_{\geq 0}^n$ converges exponentially fast towards the equilibrium $a \in \mathbb{R}_{>0}^n$ of the ODE system $\dot{x} = Ax$ starting at $x(0) \in \mathbb{R}_{\geq 0}^n$, that is, there exists $\gamma > 0$ and $K > 0$ such that

$$\| x(t) - a \| \leq K e^{-\gamma t} \quad \text{for} \quad t \geq 0.$$

Proof. We will first rephrase the ODE system such that standard theory is applicable. Let rank of A be $n - r$. Let W be as defined in Lemma 1, that is, W be an $r \times n$ matrix comprising of r linearly independent left kernel vectors of A so that $WA = 0$. Here since rank of A is $n - r$, the rows of W would form a basis for the left kernel of A. And as in Lemma 1, further suppose W is in the row reduced form, that is,

$$W = \left(W' \; I_r \right).$$

Then

$$\dot{x} = Ax \tag{3}$$

is a linear dynamical system with r conservation laws (one for each row of W). Let $Wx(0) = T \in \mathbb{R}^r$ be the vector of conserved amounts. Let $\hat{x} = (x_1, \ldots, x_{n-r})$ and $\tilde{x} = (x_{n-r+1}, \ldots, x_n)$. We will consider the (equivalent) dynamical system in which r variables are eliminated, expressed through the conservation laws

$$T = Wx = \left(W' \; I_r \right) x, \quad \text{or} \quad \tilde{x} = T - W'\hat{x}.$$

As in Lemma 1, let A be given as

$$A = \begin{pmatrix} A_{11} & A_{12} \\ A_{21} & A_{22} \end{pmatrix},$$

where A_{11} is a $(n - r) \times (n - r)$ matrix, A_{12} is a $(n - r) \times r$ matrix, A_{21} is a $r \times (n - r)$ matrix, and A_{22} is a $r \times r$ matrix. This yields

$$
\begin{aligned}
\dot{\hat{x}} &= \left(A_{11} \; A_{12} \right) \begin{pmatrix} \hat{x} \\ \tilde{x} \end{pmatrix} = \left(A_{11} \; A_{12} \right) \begin{pmatrix} \hat{x} \\ T - W'\hat{x} \end{pmatrix} \\
&= (A_{11} - A_{12}W')\hat{x} + A_{12}T \\
&= C\hat{x} + DT,
\end{aligned}
\tag{4}
$$

with $C = A_{11} - A_{12}W'$ and $D = A_{12}$. We call this as the *reduced ODE system*. Note that this reduced system has only $n - r$ variables and that the conservation laws are built directly into it. This implies that the differential equation changes if T is changed. The role of this construction is so that we can work only with the non-zero eigenvalues of the A.

As we also have $WA(t) = 0$ for all $t \geq 0$, the ODE $\dot{x} = A(t)x$ can also be similarly reduced to

$$\dot{\hat{x}} = C(t)\hat{x} + D(t)T, \tag{5}$$

with $C(t) = A_{11}(t) - A_{12}(t)W'$ and $D(t) = A_{12}(t)$, where analogous to A_{11}, we define $A_{11}(t)$ to be the top-left $(n-r) \times (n-r)$ sub-matrix of $A(t)$ and analogous to A_{12}, we define $A_{12}(t)$ to be the top-right $(n - r) \times r$ sub-matrix of $A(t)$.

Now if a is the equilibrium of the ODE system $\dot{x} = Ax$ starting at $x(0)$, then $\hat{a} = (a_1, \ldots, a_{n-r})$ is an equilibrium of the reduced ODE system $\dot{\hat{x}} = C\hat{x} + DT$ starting at $\hat{x}(0) = (x_1(0), \ldots, x_{n-r}(0))$. Suppose we are able to prove that solutions of reduced ODE $\dot{\hat{x}} = C(t)\hat{x} + D(t)T$ starting at $\hat{x}(0)$ converges exponentially fast towards $\hat{a}$ then because of the conservation laws $\tilde{x} = T - W'\hat{x}$, we would also have that solutions of $\dot{x} = A(t)x$ starting at $x(0)$ converges exponentially fast towards a. So henceforth, we will work only with the reduced

ODE systems. For notational convenience, we will drop the hats off $\hat{x}$ and $\hat{a}$ and simply refer to them as x and a respectively.

By subtracting and adding terms to the reduced ODE system (in Eq. 5), we have

$$\dot{x} = C(t)x + D(t)T$$
$$= Cx + DT + (C(t) - C)x + (D(t) - D)T.$$

As a is an equilibrium of the ODE system $\dot{x} = Cx + DT$, we have $Ca + DT = 0$. Define $y = x - a$. Then

$$\dot{y} = Cx + DT + (C(t) - C)x + (D(t) - D)T$$
$$= Cy + Ca + DT + (C(t) - C)x + (D(t) - D)T$$
$$= Cy + (C(t) - C)x(t) + (D(t) - D)T$$
$$= Cy + E(t)$$

where it is used that $Ca + DT = 0$, and $E(t) = (C(t) - C)x(t) + (D(t) - D)T$. The solution to the above ODE system is known to be

$$y(t) = e^{Ct}y(0) + \int_0^t e^{C(t-s)} E(s) \, ds. \tag{6}$$

We have, using (6) and the triangle inequality,

$$\| y(t) \| \leq \| e^{Ct} y(0) \| + \int_0^t \| e^{C(t-s)} \| \, \| E(s) \| \, ds.$$

Now A as defined (see Eq. 1 with k inserted for $k(t)$) would form a Laplacian matrix over a strongly connected graph and so it follows that all the eigenvalues of A are either zero or have negative real part. And using $C = A_{11} - A_{12}W'$ and Lemma 1 it follows that all eigenvalues of C have negative real part. Hence it follows that

$$\| e^{Ct} \| \leq K_2 e^{-\gamma_2 t},$$

where $0 < \gamma_2 < -\Re(\lambda_1)$ and $K_2 > 0$. Here λ_1 is the eigenvalue of C with the largest real part.

The matrices $C(t)$ and $D(t)$ are linear in $k(t)$. And as $k(t)$ converges exponentially fast towards k, it follows that the matrices $C(t)$ and $D(t)$ converge exponentially fast towards C and D respectively. Hence it follows that

$$\| E(t) \| = \| (C(t) - C)x(t) + (D(t) - D)T \|$$
$$\leq \| (C(t) - C) \| \, \| x(t) \| + \| (D(t) - D) \| \, \| T \|$$
$$\leq K_3 e^{-\gamma_3 t} + K_4 e^{-\gamma_4 t}$$
$$\leq K_5 e^{-\gamma_5 t}$$

where

- $\|(C_0(t) - C)\| \, \|x(t)\| \leq K_3 e^{-\gamma_3 t}$ for some $K_3, \gamma_3 > 0$ as $C(t)$ converges exponentially fast towards C and $x(t)$ is bounded (as in the original ODE $\sum_{i=1}^{n} x_i$ is conserved), and
- $\|(D_0(t) - D)\| \, \|T\| \leq K_4 e^{-\gamma_4 t}$ for some $K_4, \gamma_4 > 0$ as $D(t)$ converges exponentially fast towards D, and
- $K_5 = \frac{1}{2} \max(K_3, K_4)$ and $\gamma_5 = \min(\gamma_3, \gamma_4)$.

Collecting all terms we have for all $t \geq 0$,

$$\|y(t)\| \leq \|y(0)\| K_2 e^{-\gamma_2 t} + \int_0^t K_2 e^{-\gamma_2(t-s)} \times K_5 e^{-\gamma_5 s} ds$$

$$\leq K_0 e^{-\gamma_0 t} + K_0 \int_0^t e^{-\gamma_0 t} ds$$

$$= K_0 e^{-\gamma_0 t}(1+t)$$

$$\leq K e^{-\gamma t}$$

by choosing $K_0 = \max(K_2 K_5, \|y(0)\| K_2)$ and $\gamma_0 = \min(\gamma_1, \gamma_2)$. In the last line γ is chosen such that $0 < \gamma < \gamma_0$ and K is sufficiently large. Since $y(t) = x(t) - a$ we have,

$$\|x(t) - a\| \leq K e^{-\gamma t},$$

as required.   □

We will now prove Theorem 3. For the sake of convenience, we first recall the statement.

Theorem 3. *For the Baum Welch Reaction System $(BW(\mathcal{M}, h^*, v^*, L), k)$ with permissible rates k, if the concentrations of θ and ψ species are held fixed at a positive point then the Forward, Backward and Expection step reaction systems on α, β, γ and ψ species converge to equilibrium exponentially fast.*

Proof. It follows by repeated use of Lemma 2. For $l = 1$ the forward reaction network can be interpreted as the following molecular reactions:

$$\alpha_{1h} \xrightarrow{\pi_{h^*} \psi_{h^* w} E_{1w}} \alpha_{1h^*}$$

$$\alpha_{1h^*} \xrightarrow{\pi_h \psi_{hw} E_{1w}} \alpha_{1h}$$

for all $h \in H \setminus \{h^*\}$ and $w \in V$, as they are dynamically equivalent. Here the effective rate constants ($\pi_{h^*} \psi_{h^* w} E_{1w}$ or $\pi_h \psi_{hw} E_{1w}$) are independent of time and so the conditions of Lemma 2 are fulfilled. Thus this portion of the reaction network converges exponentially fast.

The rest of the forward reaction network can be similarly interpreted as the following molecular reactions:

$$\alpha_{lh} \xrightarrow{\alpha_{l-1,g} \theta_{gh^*} \psi_{h^* w} E_{lw}} \alpha_{lh^*}$$

$$\alpha_{lh^*} \xrightarrow{\alpha_{l-1,g} \theta_{gh} \psi_{hw} E_{lw}} \alpha_{lh}$$

for all $g \in H$, $h \in H \setminus \{h^*\}$, $l = 2, \ldots, L$ and $w \in V$. For layers $l = 2, \ldots, L$ of the forward reaction network we observe that the effective rate constants ($\alpha_{l-1,g} \theta_{gh^*} \psi_{h^*w} E_{lw}$ or $\alpha_{l-1,g} \theta_{gh} \psi_{hw} E_{lw}$) for layer l depend on time only through $\alpha_{l-1,g}$. If we suppose that the concentration of $\alpha_{l-1,g}$ converges exponentially fast, then we can use Lemma 2 to conclude that the concentration of α_{lh} also converges exponentially fast. Thus using Lemma 2 inductively layer by layer we conclude that forward reaction network converges exponentially fast. The backward reaction network converges exponentially fast, similarly.

For the expectation reaction network it likewise follows by induction. But here, notice if we interpreted the expectation network similarly into molecular reactions, the effective rate constants would depend on time through the products such as $\alpha_{lh} \beta_{lh}$ or $\alpha_{lh} \beta_{l+1,h}$. So to apply Lemma 2 we need the following: If $\alpha_l(t)$ and $\beta_l(t)$ converge exponentially fast towards a_l and b_l then the product $\alpha_l(t)\beta_l(t)$ converges exponentially fast towards $a_l b_l$ as

$$\| \alpha_l(t)\beta_l(t) - a_l b_l \| = \| (\alpha_l(t) - a_l)(\beta_l(t) - b_l) + a_l(\beta_l(t) - b_l) + b_l(\alpha_l(t) - a_l) \|$$
$$\leq \| \alpha_l(t) - a_l \| \, \| \beta_l(t) - b_l \| + K \| \beta_l(t) - b_l \| + K \| \alpha_l(t) - a_l \|,$$

where K is some suitably large constant. We can further observe that $\alpha_l(t)\beta_l(t)$ converges exponentially fast towards $a_l b_l$ at rate $\gamma = \min(\gamma_a, \gamma_b)$, where γ_a and γ_b, respectively, are the exponential convergence rates of $\alpha_l(t)$ and $\beta_l(t)$. A similar argument goes for the products of the form $\alpha_l(t)\beta_{l+1}(t)$. And thus the expectation reaction network, also converges exponentially fast.   □

We will now prove Theorem 4. For the sake of convenience, we first recall the statement.

Theorem 4. *For the Baum Welch Reaction System $(BW(\mathcal{M}, h^*, v^*, L), k)$ with permissible rates k, if the concentrations of α, β, γ and ξ species are held fixed at a positive point then the Maximization step reaction system on θ and ψ converges to equilibrium exponentially fast.*

Proof. Exponential convergence of the maximisation network follows by a similar layer by layer inductive use of Lemma 2.   □

References

1. Soloveichik, D., Seelig, G., Winfree, E.: DNA as a universal substrate for chemical kinetics. Proc. Natl. Acad. Sci. **107**(12), 5393–5398 (2010)
2. Srinivas, N.: Programming chemical kinetics: engineering dynamic reaction networks with DNA strand displacement. Ph.D. thesis, California Institute of Technology (2015)
3. Qian, L., Soloveichik, D., Winfree, E.: Efficient turing-universal computation with DNA polymers. In: Sakakibara, Y., Mi, Y. (eds.) DNA 2010. LNCS, vol. 6518, pp. 123–140. Springer, Heidelberg (2011). https://doi.org/10.1007/978-3-642-18305-8_12
4. Cardelli, L.: Strand algebras for DNA computing. Nat. Comput. **10**, 407–428 (2011)
5. Lakin, M.R., Youssef, S., Cardelli, L., Phillips, A.: Abstractions for DNA circuit design. J. R. Soc. Interface **9**(68), 470–486 (2011)

6. Cardelli, L.: Two-domain DNA strand displacement. Math. Struct. Comput. Sci. **23**, 02 (2013)

7. Chen, Y.-J., et al.: Programmable chemical controllers made from DNA. Nat. Nanotechnol. **8**(10), 755–762 (2013)

8. Lakin, M.R., Stefanovic, D., Phillips, A.: Modular verification of chemical reaction network encodings via serializability analysis. Theor. Comput. Sci. **632**, 21–42 (2016)

9. Srinivas, N., Parkin, J., Seelig, G., Winfree, E., Soloveichik, D.: Enzyme-free nucleic acid dynamical systems. Science **358**, 6369 (2017)

10. Cherry, K.M., Qian, L.: Scaling up molecular pattern recognition with DNA-based winner-take-all neural networks. Nature **559**, 7714 (2018)

11. Zechner, C., Seelig, G., Rullan, M., Khammash, M.: Molecular circuits for dynamic noise filtering. Proc. Natl. Acad. Sci. **113**(17), 4729–4734 (2016)

12. Badelt, S., Shin, S.W., Johnson, R.F., Dong, Q., Thachuk, C., Winfree, E.: A general-purpose CRN-to-DSD compiler with formal verification, optimization, and simulation capabilities. In: Brijder, R., Qian, L. (eds.) DNA 2017. LNCS, vol. 10467, pp. 232–248. Springer, Cham (2017). https://doi.org/10.1007/978-3-319-66799-7_15

13. Lakin, M.R., Youssef, S., Polo, F., Emmott, S., Phillips, A.: Visual DSD: a design and analysis tool for DNA strand displacement systems. Bioinformatics **27**(22), 3211–3213 (2011)

14. Hjelmfelt, A., Weinberger, E.D., Ross, J.: Chemical implementation of neural networks and Turing machines. Proc. Natl. Acad. Sci. **88**(24), 10983–10987 (1991)

15. Buisman, H.J., ten Eikelder, H.M.M., Hilbers, P.A.J., Liekens, A.M.L.: Computing algebraic functions with biochemical reaction networks. Artif. Life **15**(1), 5–19 (2009)

16. Oishi, K., Klavins, E.: Biomolecular implementation of linear I/O systems. IET Syst. Biol. **5**(4), 252–260 (2011)

17. Soloveichik, D., Cook, M., Winfree, E., Bruck, J.: Computation with finite stochastic chemical reaction networks. Nat. Comput. **7**(4), 615–633 (2008)

18. Chen, H.-L., Doty, D., Soloveichik, D.: Deterministic function computation with chemical reaction networks. Nat. Comput. **13**(4), 517–534 (2014)

19. Qian, L., Winfree, E.: Scaling up digital circuit computation with DNA strand displacement cascades. Science **332**(6034), 1196–1201 (2011)

20. Napp, N.E., Adams, R.P.: Message passing inference with chemical reaction networks. In: Advances in Neural Information Processing Systems, pp. 2247–2255 (2013)

21. Qian, L., Winfree, E., Bruck, J.: Neural network computation with DNA strand displacement cascades. Nature **475**(7356), 368–372 (2011)

22. Cardelli, L., Kwiatkowska, M., Whitby, M.: Chemical reaction network designs for asynchronous logic circuits. Nat. Comput. **17**(1), 109–130 (2018)

23. Gopalkrishnan, M.: A scheme for molecular computation of maximum likelihood estimators for log-linear models. In: Rondelez, Y., Woods, D. (eds.) DNA 2016. LNCS, vol. 9818, pp. 3–18. Springer, Cham (2016). https://doi.org/10.1007/978-3-319-43994-5_1

24. Virinchi, M.V., Behera, A., Gopalkrishnan, M.: A stochastic molecular scheme for an artificial cell to infer its environment from partial observations. In: Brijder, R., Qian, L. (eds.) DNA 2017. LNCS, vol. 10467, pp. 82–97. Springer, Cham (2017). https://doi.org/10.1007/978-3-319-66799-7_6

25. Viswa Virinchi, M., Behera, A., Gopalkrishnan, M.: A reaction network scheme which implements the EM algorithm. In: Doty, D., Dietz, H. (eds.) DNA 2018. LNCS, vol. 11145, pp. 189–207. Springer, Cham (2018). https://doi.org/10.1007/978-3-030-00030-1_12

26. Amari, S.: Information Geometry and Its Applications. Springer, Tokyo (2016). https://doi.org/10.1007/978-4-431-55978-8

27. Csiszár, I., Matus, F.: Information projections revisited. IEEE Trans. Inf. Theor. **49**(6), 1474–1490 (2003)

28. Jaynes, E.T.: Information theory and statistical mechanics. Phys. Rev. **106**, 4 (1957)

29. Shin, J.-S., Pierce, N.A.: A synthetic DNA walker for molecular transport. J. Am. Chem. Soc. **126**(35), 10834–10835 (2004)

30. Reif, J.: The design of autonomous DNA nano-mechanical devices: walking and rolling DNA. In: DNA Computing, pp. 439–461 (2003)

31. Sherman, W., Seeman, N.: A precisely controlled DNA biped walking device. Nano Lett. **4**, 1203–1207 (2004)

32. Cover, T.M., Thomas, J.A.: Elements of Information Theory. Wiley, New York (2012)

33. Rabiner, L.R.: A tutorial on hidden Markov models and selected applications in speech recognition. Proc. IEEE **77**(2), 257–286 (1989)

34. Juang, B.H., Rabiner, L.R.: Hidden Markov models for speech recognition. Technometrics **33**(3), 251–272 (1991)

35. Feinberg, M.: On chemical kinetics of a certain class. Arch. Rational Mech. Anal **46**, 1–41 (1972)

36. Horn, F.J.M.: Necessary and sufficient conditions for complex balancing in chemical kinetics. Arch. Ration. Mech. Anal. **49**, 172–186 (1972)

37. Feinberg, M.: Lectures on chemical reaction networks (1979). http://www.che.eng.ohio-state.edu/FEINBERG/LecturesOnReactionNetworks/

38. Gopalkrishnan, M.: Catalysis in reaction networks. Bull. Math. Biol. **73**(12), 2962–2982 (2011)

39. Anderson, D.F., Craciun, G., Kurtz, T.G.: Product-form stationary distributions for deficiency zero chemical reaction networks. Bull. Math. Biol. **72**(8), 1947–1970 (2010)

40. Tu, B.P., McKnight, S.L.: Metabolic cycles as an underlying basis of biological oscillations. Nat. Rev. Mol. Cell Biol. **7**, 9 (2006)

41. McLachlan, G., Krishnan, T.: The EM Algorithm and Extensions, vol. 382. Wiley, Hoboken (2007)

42. Singh, A., Gopalkrishnan, M.: EM algorithm with DNA molecules. In: Poster Presentations of the 24th Edition of International Conference on DNA Computing and Molecular Programming (2018)

43. Poole, W., et al.: Chemical Boltzmann machines. In: Brijder, R., Qian, L. (eds.) DNA 2017. LNCS, vol. 10467, pp. 210–231. Springer, Cham (2017). https://doi.org/10.1007/978-3-319-66799-7_14

44. Roweis, S., Ghahramani, Z.: A unifying review of linear Gaussian models. Neural Comput. **11**(2), 305–345 (1999)

45. Maass, W.: On the computational power of winner-take-all. Neural Comput. **12**(11), 2519–2535 (2000)

46. Kappel, D., Nessler, B., Maass, W.: STDP installs in winner-take-all circuits an online approximation to hidden Markov model learning. PLoS Comput. Biol. **10**, 3 (2014)

Efficient Parameter Estimation for DNA Kinetics Modeled as Continuous-Time Markov Chains

Sedigheh Zolaktaf[1(✉)], Frits Dannenberg[2], Erik Winfree[2],
Alexandre Bouchard-Côté[1], Mark Schmidt[1], and Anne Condon[1]

[1] University of British Columbia, Vancouver, BC, Canada
nasimzf@cs.ubc.ca
[2] California Institute of Technology, Pasadena, CA, USA

Abstract. Nucleic acid kinetic simulators aim to predict the kinetics of interacting nucleic acid strands. Many simulators model the kinetics of interacting nucleic acid strands as continuous-time Markov chains (CTMCs). States of the CTMCs represent a collection of secondary structures, and transitions between the states correspond to the forming or breaking of base pairs and are determined by a nucleic acid kinetic model. The number of states these CTMCs can form may be exponentially large in the length of the strands, making two important tasks challenging, namely, mean first passage time (MFPT) estimation and parameter estimation for kinetic models based on MFPTs. Gillespie's stochastic simulation algorithm (SSA) is widely used to analyze nucleic acid folding kinetics, but could be computationally expensive for reactions whose CTMC has a large state space or for slow reactions. It could also be expensive for arbitrary parameter sets that occur in parameter estimation. Our work addresses these two challenging tasks, in the full state space of all non-pseudoknotted secondary structures of each reaction. In the first task, we show how to use a reduced variance stochastic simulation algorithm (RVSSA), which is adapted from SSA, to estimate the MFPT of a reaction's CTMC. In the second task, we estimate model parameters based on MFPTs. To this end, first, we show how to use a generalized method of moments (GMM) approach, where we minimize a squared norm of moment functions that we formulate based on experimental and estimated MFPTs. Second, to speed up parameter estimation, we introduce a fixed path ensemble inference (FPEI) approach, that we adapt from RVSSA. We implement and evaluate RVSSA and FPEI using the Multistrand kinetic simulator. In our experiments on a dataset of DNA reactions, FPEI speeds up parameter estimation compared to inference using SSA, by more than a factor of three for slow reactions. Also, for reactions with large state spaces, it speeds up parameter estimation by more than a factor of two.

© Springer Nature Switzerland AG 2019
C. Thachuk and Y. Liu (Eds.): DNA 25, LNCS 11648, pp. 80–99, 2019.
https://doi.org/10.1007/978-3-030-26807-7_5

1   Introduction

Nucleic acid kinetic simulators [9,29,34,35] aim to predict the kinetics of interacting nucleic acid strands, such as the rate of a reaction or the sequence of interactions between the strands. These simulators are desirable for building nucleic
acid-based devices whose nucleic acid sequences need to be carefully designed to
control their behaviour. For example, neural networks can be realized in DNA
using strand displacement reactions [6]. However, the rates of reactions vary by
several orders of magnitude depending on sequence and conditions and are hard
to predict, making the design of artifacts challenging. Accurate kinetic simulators would allow many, though not all, unanticipated design flaws to be identified
prior to conducting wet-lab experiments, and would allow more complex molecular devices to be designed and successfully implemented with fewer deficiencies
needing to be debugged experimentally.

Because of these pressing needs, there has been great progress on simulators that can model the kinetics of interacting nucleic acid strands. The simulators range from coarse-grained models that consider large rearrangements
of the base pairs [34,35], and often factor in tertiary structure, to elementary
step models that consider the forming or breaking of a single base pair [9,29],
and to molecular dynamics models that follow the three-dimensional motion of
the polymer chains [27,31]. Elementary step models are of interest to us here
because they are computationally more efficient than molecular dynamics, yet
they also can represent and thus discover unexpected sequence-dependent secondary structures within intermediate states. Continuous-time Markov chains
(CTMCs) play a central role in modeling nucleic acid kinetics with elementary
steps, such as Kinfold [9] and Multistrand [28,29]. States of the CTMCs correspond to secondary structures and have exponentially distributed holding times,
and transitions between states correspond to forming or breaking of a single
base pair. Nucleic acid kinetic models [24,42], along with nucleic acid thermal
stability models [2,17,38,39], specify the rate of transition between states and
the holding time of states. These simulators can stochastically sample paths
(sequences of states from an initial to a target state) and trajectories (sequences
of states from an initial to a target state, along with the times to transition
between successive states). The mean first passage time (MFPT) from an initial
to a target state can be estimated from sampled trajectories. The first passage
time of a trajectory is the first time that the trajectory occupies the target state.
Kinetic rates, such as the rate constant of a reaction [28], can then be derived
from such estimates.

Our work addresses two challenging tasks in accurately predicting the MFPT
of a reaction's CTMC, in the full state space of all non-pseudoknotted[1] secondary
structures. The first task is to estimate the MFPT of a reaction's CTMC, given
a calibrated kinetic model. The second task is to calibrate parameters of kinetic

---

[1] A pseudoknot is a secondary structure that has at least two base pairs in which one
nucleotide of a base pair is intercalated between the two nucleotides of the other
base pair.

models; even though thermal stability models are well calibrated [3,23], parameters of kinetic models, which affect the rate of transition between states and consequently holding times of states, are not well calibrated [42]. These tasks are challenging, particularly for multistranded DNA kinetics, because when nucleic acid strands interact, they are prone to the formation of many metastable secondary structures due to stochastic formation and breakage of base pairs. The number of possible secondary structures nucleic acids can form may be exponentially large compared to the number of nucleotides the strands contain. Moreover, to make accurate estimations, many sampled trajectories might be required, which might be time-consuming to obtain (see Sect. 4). In this work, we make progress on these tasks, by focusing on the Multistrand kinetic simulator [28,29] (described in Sect. 2.1), that is used to analyze the folding kinetics of multiple interacting nucleic acid strands and models the kinetics as CTMCs with elementary steps. In the rest of this section, first, we describe related work for MFPT estimation and our contributions. Then, we describe related work for calibrating kinetic models based on MFPTs and our contributions.

1.1   Mean First Passage Time Estimation

Exact linear algebra methods [33,42] can provide an exact solution to the MFPT of a CTMC that can be explicitly represented. However, their accuracy could be compromised by numerical errors and it is infeasible to use these methods for CTMCs with large implicitly-represented state spaces. Our previous work [42] estimates MFPTs on heuristically obtained reduced state spaces of the CTMCs. Moreover, the state spaces are customized for each type of reaction. In contrast to exact linear algebra methods, the MFPT could be approximated in the full state space or reduced state space with Gillespie's stochastic simulation algorithm (SSA) [11]. SSA can be slow depending on the CTMC of the reaction. We could adapt sequential Monte Carlo and importance sampling techniques [8,13], but these methods require a proposal distribution that efficiently reduces the variance of the estimator. More recently, machine learning algorithms have been developed to successfully predict DNA hybridization rates [41] from sequence, without enumerating the state space of the reaction. However, these methods can not treat other reactions or kinetics.

Our Contribution. We show how to use a reduced variance stochastic simulation algorithm (RVSSA), a Rao-Blackwellised version [20] of SSA, to estimate the MFPT of a reaction's CTMC. In SSA, the variance of MFPT estimates arises for two reasons. First, the path to a target state affects the MFPT. Second, the holding time in each state affects the MFPT. RVSSA removes the stochasticity in the holding times by using expected holding times of states. We prove that RVSSA produces a lower variance estimator of the MFPT compared to SSA. Moreover, we show in our experiments that RVSSA has a lower variance than SSA in estimating the MFPT of a reaction's CTMC, when in the sampled paths there exist a small number of states that have large expected holding times compared to other states. One interesting example that we identify is the association of poly(A) and poly(T) sequences in low concentrations (see Fig. 2b).

1.2   Parameter Estimation

In order to make accurate MFPT estimations, the underlying parameters of the CTMCs should be calibrated. Models of nucleic acid thermal stability [2, 17, 38, 39] have been extensively calibrated to experimental data [3, 23]. However, extensive calibration of nucleic acid kinetic models remains challenging [42]. Our previous work [42] uses a maximum a posteriori approach and a Markov chain Monte Carlo approach to calibrate DNA kinetic models on a wide range of reactions, such as strand displacement [40], but on reduced state spaces of the reactions. The reduced state spaces are manually designed and coded and the largest reduced state space contains less than 1.5×10^4 states. Moreover, related work [32, 40] uses reaction specific models to calibrate a kinetic model. These models are not easily adapted to other kinetic models. There have been advances in calibrating CTMCs [1, 21, 30] based on transient probabilities, the likelihood that a process will be in a given state at a given time, and these advances have been used for calibration of nucleic acid kinetics [13] and chemical reaction networks [10, 12, 19, 22].

During the optimization, for every new parameter set, we could use SSA or RVSSA to obtain an unbiased estimate of the MFPT of a reaction's CTMC. However, sampling new trajectories for every new parameter set could be computationally expensive for large CTMCs, slow reactions, or arbitrary parameter sets. One reason is that transitions might be repeatedly sampled. We could also use importance sampling techniques [8, 13], but these methods would require a proposal distribution that efficiently reduces the variance of the estimator, which is challenging when the underlying transition probabilities of the CTMC are changing throughout parameter estimation.

Our Contribution. To estimate parameters for DNA kinetics modeled as CTMCs based on MFPTs, we show how to use a generalized method of moments (GMM) [14] estimator. More importantly, we show how to use a fixed path ensemble inference (FPEI) approach that speeds up parameter estimation compared to a reference method that uses SSA directly during inference (SSAI). The GMM method is widely used in econometrics and has also recently been used in chemical reaction networks [22]. The GMM method can be used when a maximum likelihood estimate or a maximum a posteriori estimate is infeasible, as is the case with CTMCs that have very large state spaces. The GMM method minimizes a weighted norm of moment conditions obtained from samples. The moment conditions are functions of model parameters and the dataset such that the expectation of the moment conditions is zero at the true value of the parameters. To minimize the squared norm of the moment conditions, we use the Nelder-Mead direct-search optimization algorithm [26], which has been frequently used in optimization problems that have small stochastic perturbations in function values [4].

To speed up parameter estimation, we introduce and use FPEI. In this method, we condense paths, where for every path, we compute the set of states and the number of times each state is visited. Rather than generating new trajectories with SSA for every parameter set variation (the SSAI method), in FPEI we use fixed condensed paths to speed up parameter estimation. For example, in this work, the length of the longest path is more than 1×10^8, whereas the number of unique states and transitions of the path is approximately 3.8×10^5 and 1.4×10^6, respectively. In FPEI, we use RVSSA to estimate the MFPT of the fixed paths given a new parameter set. Since the MFPT estimates obtained with fixed paths are biased, we alternate between minimizing the error of prediction on fixed paths, and resampling new paths and restarting the optimization method.

To implement RVSSA and FPEI, we augment the Multistrand kinetic simulator [28,29] where for each reaction the full state space of all non-pseudoknotted secondary structures is possible. We conduct computational experiments on experimental DNA reactions that have moderate or large state spaces or are slow, including hairpin closing, hairpin opening, helix dissociation with and without mismatches, and helix association. We compare the performance of RVSSA with SSA for MFPT estimation and FPEI with SSAI for parameter estimation. Results for our example data are encouraging, showing that FPEI speeds up parameter estimation compared to using SSAI, by more than a factor of three for slow reactions. Also, for reactions with large state spaces, it speeds up parameter estimation by more than a factor of two.

2   Preliminaries

2.1   The Multistrand Kinetic Simulator

The Multistrand kinetic simulator [28,29] models the kinetics of multiple interacting nucleic acid strands as a CTMC. A state of the CTMC represents a system microstate, in other words, the configuration of the strands in the fixed volume that we simulate. A system microstate is a collection of complex microstates. A complex microstate is a set of strands connected by base pairing (secondary structures). In Multistrand, all possible secondary structures are permitted except for pseudoknots. Multistrand defines the energy of a state as the sum of the standard free energy for each complex, which is determined with Multistrand's nucleic acid thermal stability model. Transitions between states correspond to the forming or breaking of a single base pair. For example, in Fig. 1, state t can transition to states s and u. The transition rate k_{ts} from state t to state s is determined by the energy of the states and a nucleic acid kinetic model.

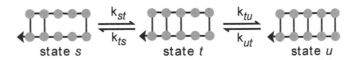

Fig. 1. State t can transition to states s and u by breaking or forming a single base pair, respectively. The reverse transitions are also possible.

We experiment with the Metropolis [24] and the Arrhenius [42] kinetic models that are implemented in the Multistrand software. The Metropolis kinetic model has two free parameters k_{uni} and k_{bi} that distinguish between unimolecular and bimolecular transitions, respectively. In the Arrhenius kinetic model, transition rates additionally depend on the local context of the base pair that is forming or breaking. The model differentiates between seven different half contexts $C = \{stack, loop, end, stack+loop, stack+end, loop+end, stack+stack\}$. For example, in Fig. 1, in the transition from state t to state s, the half contexts of the base pair that is breaking are a stack and a loop. An Arrhenius rate constant A_l and an activation energy E_l are associated with each half context l. The model also has a bimolecular scaling constant α. In total, the model has 15 free parameters.

To sample paths and trajectories for a reaction, experimental conditions need to be determined, such as the sequence of the strands, the temperature, the concentration of Na^+ and Mg^{2+} cations, and the initial concentration of the strands. We adopt the trajectory mode of Multistrand for all reactions of our dataset. In this mode, SSA is implemented to simulate trajectories over the CTMC of a reaction, starting in an initial state and halting when the reaction is over, and to estimate the MFPT. For helix association and hairpin closing reactions, all trajectories start from the state where no base pairs are formed and end at the state where the duplex is fully formed. For hairpin opening and helix dissociation the start and end states are reversed. Given the estimated MFPT $\hat{\tau}^r$ of reaction r, as computed over several trajectories, the reaction rate constant of reaction r is computed as

$$\hat{k}^r = \begin{cases} \frac{1}{\hat{\tau}^r} & \text{first order reaction} \\ \frac{1}{\hat{\tau}^r u^r} & \text{second order reaction} \end{cases}, \tag{1}$$

where u^r is the initial concentration of the reactants of reaction r in the simulation [28]. Equation (1) also holds for the experimental reaction rate constant and the experimental MFPT, called timescale, of the reaction.

2.2 Gillespie's Stochastic Simulation Algorithm

Gillespie's stochastic simulation algorithm (SSA) [11] has been widely used to simulate stochastic trajectories in CTMCs [28,29]. It provides an unbiased and consistent estimate of the MFPT from an initial state to a target state. It estimates the MFPT as the mean of the first passage times of sampled trajectories. In brief, to sample a trajectory and its first passage time, SSA advances forward in two steps:

1. At a jump from the current state s_i, SSA samples the holding time T_i of the state from an exponential distribution with a rate equal to the sum of the transition rates from the state, in other words, $T_i \mid s_i \sim \text{Exp}(k_{s_i})$, where $k_{s_i} = \sum_{s \in S} k_{s_i s}$, S is the state space of the CTMC, $k_{s_i s}$ is the transition rate from state s_i to state s, if s is not a neighbor of s_i then $k_{s_i s} = 0$, $\mathbb{E}[T_i \mid s_i] = k_{s_i}^{-1}$ and $\text{Var}(T_i \mid s_i) = k_{s_i}^{-2}$.

2. At a jump from the current state s_i, SSA samples the next state s_{i+1} from the outgoing transition probabilities of state s_i, in other words, $p(s_i, s) = \frac{k_{s_i s}}{k_{s_i}}$, $s_i \neq s$.

Let P be a trajectory of length Z from state s to state t, with holding times $T_1, ..., T_{Z-1}$, obtained by using SSA with initial state s, and ending the first time that state t is sampled. In SSA, the FPT of the trajectory is computed as

$$F^{\text{SSA}} = \sum_{i=1}^{Z-1} T_i. \tag{2}$$

By using N independently sampled trajectories, we obtain a Monte Carlo estimator for the MFPT of the CTMC as $\hat{\tau}_N^{\text{SSA}} = \frac{1}{N} \sum_{n=1}^{N} F_n^{\text{SSA}}$.

3   Methodology

3.1   Mean First Passage Time Estimation

In SSA, the variance of MFPT estimates arises for two reasons. First, the path to a target state affects the MFPT. Second, the holding time in each state affects the MFPT. Hordijk et al. [18] show how to obtain a reduced variance estimate of a steady-state measure of an irreducible and positive recurrent CTMC. Their constant holding-time method eliminates the variability in the random holding time of states and instead uses expected holding times. To estimate the MFPT of a reaction's CTMC, we formulate a Rao-Blackwellised version [20] of SSA, which similar to Hordijk et al. also eliminates the variability in the random holding times of states. However, the CTMC is not restricted to be irreducible or positive recurrent and the MFPT estimate is not necessarily a steady-state measure. We call this method the reduced variance stochastic simulation algorithm (RVSSA). Similar to SSA, RVSSA also produces a consistent and unbiased estimator of the MFPT, but has a smaller variance in predicting MFPTs compared to SSA[2].

In brief, in RVSSA, instead of sampling a random holding time for each state, we use an estimator based on the expected holding time. The algorithm proceeds as follows.

1. At a jump from the current state s_i, compute the expected holding time $\overline{T}_i$ before jumping to the next state, in other words, $\overline{T}_i = k_{s_i}^{-1} = (\sum_{s \in \mathcal{S}} k_{s_i s})^{-1}$. Note that $\mathbb{E}[\overline{T}_i \mid s_i] = k_{s_i}^{-1}$ and $\text{Var}(\overline{T}_i \mid s_i) = 0$.
2. Step 2 of SSA.

---

[2] For our purpose here, we are only interested in the MFPT, so the smaller variance is good. In other contexts, the full distribution of FPTs will be of interest, and for that purpose only SSA, but not RVSSA, will be appropriate.

Let P be a path of length Z from state s to state t, with expected holding times $\overline{T}_1, ..., \overline{T}_{Z-1}$, obtained by using RVSSA with initial state s, and ending the first time that state t is sampled. In RVSSA, we compute the MFPT of the path as

$$Y^{\text{RVSSA}} = \sum_{i=1}^{Z-1} \overline{T}_i. \tag{3}$$

By using N independently sampled paths, we obtain a Monte Carlo estimator for the MFPT of the CTMC as $\hat{\tau}_N^{\text{RVSSA}} = \frac{1}{N} \sum_{n=1}^{N} Y_n^{\text{RVSSA}}$.

Theorem 1. *The estimator of the MFPT from state s to state t produced by RVSSA has a lower variance than the estimator produced by SSA.*

Proof. Let P denote a random path from state s to state t. We have $\mathbb{E}[F^{\text{SSA}} \mid P] = \mathbb{E}[Y^{\text{RVSSA}} \mid P]$, and consequently

$$\text{Var}(\mathbb{E}[F^{\text{SSA}} \mid P]) = \text{Var}(\mathbb{E}[Y^{\text{RVSSA}} \mid P]). \tag{4}$$

Also, $\mathbb{E}[\text{Var}(F^{\text{SSA}} \mid P)] > 0$, and $\mathbb{E}[\text{Var}(Y^{\text{RVSSA}} \mid P)] = \mathbb{E}[\text{Var}(\sum_{i=1}^{Z-1} \overline{T}_i \mid P)] = 0$ because $\overline{T}_i$ are constants and independent given P. Based on the law of total variance

$$\text{Var}(Y^{\text{RVSSA}}) = \mathbb{E}[\text{Var}(Y^{\text{RVSSA}} \mid P)] + \text{Var}(\mathbb{E}[Y^{\text{RVSSA}} \mid P]) \overset{\text{by Eq. (4)}}{=}$$

$$\mathbb{E}[\text{Var}(Y^{\text{RVSSA}} \mid P)] + \text{Var}(\mathbb{E}[F^{\text{SSA}} \mid P]) = \text{Var}(\mathbb{E}[F^{\text{SSA}} \mid P]) < \tag{5}$$

$$\mathbb{E}[\text{Var}(F^{\text{SSA}} \mid P)] + \text{Var}(\mathbb{E}[F^{\text{SSA}} \mid P]) = \text{Var}(F^{\text{SSA}}).$$

Therefore, it can be concluded that $\text{Var}(\hat{\tau}_N^{\text{RVSSA}}) = \text{Var}(\frac{1}{N} \sum_{n=1}^{N} Y_N^{\text{RVSSA}}) = \frac{1}{N} \text{Var}(Y^{\text{RVSSA}}) < \frac{1}{N} \text{Var}(F^{\text{SSA}}) = \text{Var}(\hat{\tau}_N^{\text{SSA}})$. $\square$

For an unbiased estimator, the expected mean squared error (MSE) of the estimator is equal to the variance of the estimator [36]. Consequently, RVSSA has a smaller MSE than SSA and requires fewer sampled paths to estimate the MFPT,

$$\mathbb{E}[(\hat{\tau}_N^{\text{RVSSA}} - \tau)^2] = \frac{1}{N} \text{Var}(Y^{\text{RVSSA}}) < \frac{1}{N} \text{Var}(F^{\text{SSA}}) = \mathbb{E}[(\hat{\tau}_N^{\text{SSA}} - \tau)^2]. \tag{6}$$

3.2 Parameter Estimation

In Sect. 3.1, we assume that the underlying parameters of the CTMCs are known. Here, we focus on estimating the underlying parameters of the transition rates when they are not known a priori.

To estimate model parameters, we formulate a generalized method of moments (GMM) [14] objective function based on experimental and predicted MFPTs. The GMM estimators have desirable statistical properties under suitable conditions, such as consistency and asymptotic normality. The GMM method minimizes a weighted norm of moment conditions. The moment conditions are functions of model parameters and observed values such that the expectation of the moment conditions is zero at the true value of the parameters. Given a column vector $\mathbf{g}$ of moment conditions and its transpose $\mathbf{g}^{\mathrm{T}}$, the GMM method seeks the true parameter set θ^* as

$$\theta^* = \operatorname*{argmin}_{\theta} \mathbf{g}(\theta)^{\mathrm{T}} \mathbf{W} \mathbf{g}(\theta), \tag{7}$$

where $\mathbf{W}$ is a positive-definite matrix that controls the variance of the estimator. For optimally chosen weights, which depend on the covariance of the moment conditions at the true parameter set θ^*, the estimator has the smallest possible variance for the parameters. Since the true parameter set is unknown, there exist several approaches to deal with this issue. For example, the two-step GMM estimator [15] uses the identity matrix in the first step to estimate a parameter set. In the second step, it uses the estimated parameters to produce the weighting matrix and reestimates the parameters. In our experiments, we only use the identity weighting matrix, which produces a consistent and asymptotic normal GMM estimator, and leave other options to future work.

Let θ be a parameter set for a kinetic model that parameterizes the CTMC of reactions, and let θ^* be the true parameter set. For reaction r, based on the experimental MFPT τ^r and an unbiased estimator of the MFPT $\hat{\tau}^r$, we can define a moment condition as $g^r(\theta) = \hat{\tau}^r(\theta) - \tau^r$. However, since reactions occur at timescales that cover many orders of magnitude, from slow reactions, such as helix dissociation, to faster reactions, such as hairpin closing, and since we are using an identity matrix, we use $\log_{10}$ differences instead; we define a moment condition as

$$g^r(\theta) = \log_{10} \hat{\tau}^r(\theta) - \log_{10} \tau^r, \tag{8}$$

where we approximate $\mathbb{E}[g^r(\theta^*)] = \mathbb{E}[\log_{10} \hat{\tau}^r(\theta^*)] - \log_{10} \tau^r \approx 0$ for the true parameter set θ^* (if one exists). This approximation is reasonable for an unbiased and low variance estimator of the experimental MFPT τ^r. The Taylor expansion of $\mathbb{E}[\log_{10} \hat{\tau}^r(\theta^*)]$ around $\log_{10} \mathbb{E}[\hat{\tau}^r(\theta^*)] = \log_{10} \tau^r$ is $\mathbb{E}[\log_{10} \hat{\tau}^r(\theta^*)] \approx$ $\mathbb{E}\left[\log_{10} \tau^r + \frac{1}{\tau^r}(\hat{\tau}^r - \tau^r) - \frac{1}{2(\tau^r)^2}(\hat{\tau}^r - \tau^r)^2\right] = \log_{10} \tau^r - \frac{\mathrm{Var}(\hat{\tau}^r(\theta^*))}{2(\tau^r)^2}$, where the second term disappears. Also, note that based on Eq. (1), instead of Eq. (8) we equivalently use $g^r(\theta) = \log_{10} \hat{\tau}^r(\theta) - \log_{10} \tau^r = \log_{10} k^r - \log_{10} \hat{k}^r(\theta)$, which is commonly used in related work [41,42]. Based on the entire reactions of the dataset $\mathcal{D}$, we define the GMM estimator as

$$\theta^* = \operatorname*{argmin}_{\theta} \sum_{r \in \mathcal{D}} \left(\log_{10} k^r - \log_{10} \hat{k}^r(\theta)\right)^2. \tag{9}$$

This can be recognized as the least mean squared error (MSE) parameter set.

In our experiments (described in Sect. 4.3), we seek a parameter set that minimizes the GMM estimator. However, we also considered using the negative of Eq. (14) from our previous work [42], where $g^r(\theta)$ is defined to be normally distributed with an unbiased mean and variance σ^2, and a small $L2$ regularization term is also defined. With this objective function, the predictive quality of the fixed path ensemble inference (FPEI) approach, which we describe later on, only slightly changes for our dataset.

To minimize the objective function, we use the Nelder-Mead direct-search optimization algorithm [26]. To approximate a local optimum parameter set θ with size $|\theta|$, the algorithm maintains a simplex of $|\theta| + 1$ parameter sets. The algorithm evaluates the function value at every parameter set of the simplex. It proceeds by attempting to replace a parameter set that has the worst function value with a parameter set reflected through the centroid of the remaining $|\theta|$ parameter sets in the simplex with expansion and contraction as needed. The algorithm uses only the ranks of the function values to determine the next parameter set, and therefore has been frequently used in optimization problems that have small stochastic perturbations in function values [4]. This robustness is essential for its use in SSAI.

In SSAI, during the optimization, to obtain an unbiased estimate of τ^r for every parameter set variation, we use SSA. However, obtaining new trajectories for every parameter set is computationally expensive. One reason is that transitions might be repeatedly sampled. Therefore the length of a trajectory could be much larger than the number of unique states and transitions of the trajectory (see Sect. 4.3). We propose to use FPEI which uses an ensemble of fixed paths, with an efficient data structure, to speed up parameter estimation. In FPEI, for every reaction, we build a fixed set of paths with an initial parameter set θ_0. For a new parameter set θ_m, we use the fixed paths to estimate the MFPT. To speed up computations, we condense paths; for every path, we compute the set of states and the number of times each state is visited. We compute the holding time of a state in a path as if the path is regenerated in the full state space. To compute the holding time of a state under a new parameter set, we need to compute the total outgoing transition rate from the state under the new parameter set. Therefore, we also store information about the outgoing neighbors of the states that affect the outgoing transition rate. Alternatively, depending on memory and storage limitations, similar to SSA and RVSSA, we could repeatedly compute the outgoing neighbors of the states on the fly. Given this data, as the parameter set is updated to θ_m, we compute the MFPT of path P according to RVSSA as

$$Y^{\mathrm{FPEI}}(\theta_m) = \sum_{i=1}^{Z-1} \overline{T}_i(\theta_m), \text{ where } \overline{T}_i(\theta_m) = \frac{1}{k_{s_i}(\theta_m)}, \tag{10}$$

where the transition rates of the CTMC depend on the parameter set θ_m and the path is obtained with θ_0. Because of the condensed representation, this formula is not literally computed, but rather a mathematically equivalent one with fewer terms is computed. Given N fixed paths obtained with θ_0, we estimate the MFPT of the CTMC that is parameterized with θ_m as $\hat{\tau}_N^{\mathrm{FPEI}}(\theta_m) = \frac{1}{N} \sum_{n=1}^{N} Y_n^{\mathrm{FPEI}}(\theta_m)$.

With fixed paths, the MFPT estimates are biased and the learned parameter set might not perform well in the full state space where other paths are possible. Therefore, to reduce the bias and to ensure that the ensemble of paths is a fair sample with respect to the optimized parameters, we alternate between minimizing the error of prediction on fixed paths, and resampling new paths and

---

Algorithm 1. SSAI

$\theta \leftarrow \theta_0$ `// Choose initial parameter set` θ_0
Initialize the simplex in the Nelder-Mead algorithm using θ and its perturbations
while <u>stopping criteria not met</u> **do**
                    `// See Section 4.3 for our stopping criteria`
    $\theta \leftarrow$ Retrieve a parameter set from the Nelder-Mead algorithm
    Update the free parameters of the kinetic model with θ
    foreach <u>reaction $r \in$ dataset $\mathcal{D}$</u> **do**
        foreach <u>n=1,2,...,N</u> **do**
            Sample a trajectory P_n using SSA and calculate its FPT using Eq. 2
        Calculate the MFPT of the reaction using the FPTs of the trajectories
    Calculate the GMM function in Eq. 9 using the MFPT of the reactions
    Update the simplex in the Nelder-Mead algorithm based on the GMM function

---

Algorithm 2. FPEI

$\theta \leftarrow \theta_0$ `// Choose initial parameter set` θ_0
while <u>stopping criteria not met</u> **do**
                    `// See Section 4.3 for our stopping criteria`
    Update the free parameters of the kinetic model with θ
    foreach <u>reaction $r \in$ dataset $\mathcal{D}$</u> **do**
        foreach <u>n=1,2,...,N</u> **do**
            Sample a path P_n using RVSSA
            Condense path P_n for the reaction
Initialize the simplex in the Nelder-Mead algorithm using θ and its perturbations
while <u>stopping criteria not met</u> **do**
    $\theta \leftarrow$ Retrieve a parameter set from the Nelder-Mead algorithm
    Update the free parameters of the kinetic model with θ
    foreach <u>reaction $r \in$ dataset $\mathcal{D}$</u> **do**
        foreach <u>n=1,2,...,N</u> **do**
            Calculate the MFPT of path P_n using Eq. 10
        Calculate the MFPT of the reaction using the MFPTs of the paths
    Calculate the GMM function in Eq. 9 using the MFPT of the reactions
    Update the simplex in the Nelder-Mead algorithm based on the GMM function

---

restarting the optimization method. An overview of our parameter estimation framework using SSAI and FPEI, with a GMM estimator and the Nelder-Mead algorithm, is given in Algorithms 1 and 2, respectively.

We also considered a normalized importance sampling approach [8], to obtain consistent estimators of the MFPTs using fixed paths. In this approach, we also compute the set of traversed transitions and how often each of those transitions occur in the path. We weigh the estimated MFPT of each path P by the relative likelihood of the path given the new and the initial parameter sets $\tilde{L}(\theta_m) = \frac{L(\theta_m)}{L(\theta_0)}$, where $L(\theta_m)$ is the likelihood of P under parameter assignment θ_m. For RVSSA, $L(\theta_m) = \prod_{i=1}^{Z-1} \frac{k_{s_i s_{i+1}}(\theta_m)}{\sum_{s \in S} k_{s_i s}(\theta_m)} e^{-\sum_{s \in S} k_{s_i s}(\theta_m) \overline{T}_i(\theta_m)}$, and we estimate the MFPT as $\hat{\tau}_N^{\text{FPEI}}(\theta_m) = \frac{1}{\sum_{n=1}^{N} \tilde{L}_n(\theta_m)} \sum_{n=1}^{N} \tilde{L}_n(\theta_m) Y_n^{\text{FPEI}}(\theta_m)$. In our experiments, this approach performed poorly, since the effective sample size of the relative likelihoods was small.

4   Experiments

To evaluate the performance of RVSSA for MFPT estimation and FPEI for parameter estimation, in the full state space of all non-pseudoknotted secondary structures of each reaction, we augment the Multistrand kinetic simulator [28, 29] and we conduct computational experiments. Our dataset and framework are available at https://github.com/DNA-and-Natural-Algorithms-Group/FPEI.

4.1   Dataset

To evaluate the performance of RVSSA and FPEI, we use 21 experimentally determined reaction rate constants published in the literature, including hairpin closing [5], hairpin opening [5], helix association [16, 37], and helix dissociation with and without mismatches [7]. The dataset is summarized in Table 1. Each reaction of the dataset is annotated with a temperature and the concentration of Na$^+$, which affect the transition rates in the kinetic models that we use.

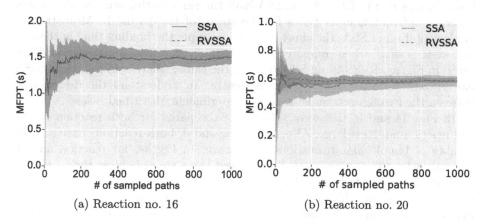

(a) Reaction no. 16   (b) Reaction no. 20

Fig. 2. The MFPT and 95% confidence interval of SSA and RVSSA, where the kinetic model is parameterized with θ_0. In both (a) and (b), RVSSA and SSA are using the same sampled paths. In (a), RVSSA and SSA have similar variance. The average computation time per sampled path, defined as the total computation time divided by the total number of sampled paths, is 3×10^2 s. In (b), RVSSA has a lower variance than SSA. The average computation time per sampled path is 0.5 s.

Table 1. Dataset of experimentally determined reaction rate constants. The concentration of the strands is set to 1×10^{-8} M, 5×10^{-8} M, and 1×10^{-8} M, for reactions no. 1–15, 16–19, and 20–21, respectively.

Reaction type & source	No.	Sequences	$T/^{\circ}$C	$[Na]^+/$M	$\log_{10} k^r$
Hairpin closing Fig. 4 from [5]	1–5	CCCAA-$(T)_{30}$-TTGGG	14.4–29.8	0.1	3.53–3.69
Hairpin opening Fig. 4 from [5]	6–10	CCCAA-$(T)_{30}$-TTGGG	14.4–29.8	0.1	2.14–3.30
Helix dissociation (with a mismatch) Fig. S4 from [7]	11–15	AGGACTTGT + ACAAGACCT AGGACTTGT + ACAAGTGCT AGGACTTGT + ACAAGTCGT AGGACTTGT + ACAAGTCCA AGGACTTGT[†]	37	0.2	0.19–0.92
Helix association Table 1 from [16]	16–19	GCCCACACTCTTACTTATCGACT[†] GCACCTCCAAATAAAAACTCCGC[†] CGTCTATTGCTTGTCACTTCCCC[†] ACCCTTTATCCTGTAACTTCCGC[†]	25	0.195	5.71–6.68
Helix association Table 1 from [37]	20–21	25-mer Poly (dA)[†] 25-mer Poly (dG)[†]	48–78	0.4	-

[†]The complement of the demonstrated sequence is also a reactant.

4.2   Mean First Passage Time Estimation

Figures 2a and b show the performance of RVSSA compared with SSA for helix association reactions no. 16 and 20, respectively. To sample paths and trajectories, we parameterize the kinetic model with the Metropolis initial parameter set [32,42], in other words, $\theta_0 = \{k_{\mathrm{uni}} = 8.2 \times 10^6\,\mathrm{s}^{-1}, k_{\mathrm{bi}} = 3.3 \times 10^5\,\mathrm{M}^{-1}\mathrm{s}^{-1}\}$. In both Figs. 2a and b, RVSSA and SSA have the same paths, but the algorithms generate different holding times for the states of the paths. In Multistrand's implementation of SSA, the effort needed to sample the holding time in the current state is small when compared to the task of computing outgoing transition rates. In Fig. 2a, RVSSA and SSA perform the same, whereas in Fig. 2b, RVSSA has a lower variance than SSA, consistently. To understand the discrepancy between the two figures, we analyze the experiments, described below.

In Figs. 3a and b, the average length of the paths for both reaction no. 16 and reaction no. 20 is large. Also, in Figs. 3c and d, both reactions have a small number of bimolecular transitions on average. In Fig. 3e, for reaction no. 16, the state where two strands are disconnected has a small holding time, because the state has many fast unimolecular transitions between complementary bases within a strand in addition to the slow bimolecular transitions. However, in Fig. 3f, for reaction no. 20, the state where the two strands are disconnected has a large holding time, since there are no complementary bases within a poly(A) or poly(T) strand and the only transitions are slow bimolecular transitions. RVSSA has a significantly lower variance for reaction no. 20 compared to SSA, because in the sampled paths, there exists a small number of states that have large expected holding times compared to other states. SSA has a large variance in generating holding times for these states. Overall, in our experiments with parameter set θ_0,

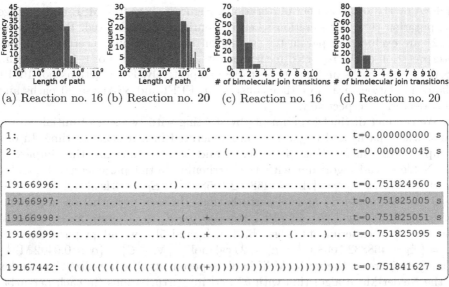

Fig. 3. Histogram of the length of 100 random paths obtained with RVSSA for (a) reaction no. 16 and (b) reaction no. 20. Histogram of the number of bimolecular join transitions of the random paths for (c) reaction no. 16 and (d) reaction no. 20. Snapshot of the i-th state visited, dot-parentheses notation and jump times for a random path obtained with RVSSA for (e) reaction no. 16 and (f) reaction no. 20. The jump time at state i is equal to the jump time at state $i-1$ plus the holding time of state $i-1$. The green highlighting indicates where a bimolecular step occurs. (Color figure online)

RVSSA has a lower variance than SSA for reactions no. 20 and 21, but performs similar to SSA for other reactions in Table 1.

4.3 Parameter Estimation

Figure 4 shows the MSE, defined as the mean of $|\log_{10} k^r - \log_{10} \hat{k}^r(\theta)|^2$ on different reactions, of FPEI and SSAI over various iterations, where the methods are learning parameters for the Arrhenius kinetic model [42]. Also, it shows the average computation time per iteration, defined as the total computation time divided by the total number of iterations. Figure 5 shows the MSE and average computation time per iteration when the entire dataset is used. Reactions no. 20–21 are excluded in parameter estimation because of our uncertainty in our interpretation of the reported measurements. For reactions no. 1–15, FPEI and SSAI use 200 paths and 200 trajectories, respectively. For reactions no. 16–19, where simulations are more time-consuming, FPEI and SSAI use 20 paths and 20 trajectories, respectively.

We conduct distinct experiments by starting with two sets of initial parameters, where paths and trajectories are generated in a reasonable time. In one group of experiments (Figs. 4a, c, e, g, and 5a), we initialize the simplex in the Nelder-Mead algorithm with the Arrhenius initial parameter set [32,42], in other words, $\theta_0' = \{A_l = 468832.1058\,\mathrm{s}^{-1/2}, E_l = 3\,\mathrm{kcal\,mol}^{-1} \mid \forall l \in \mathcal{C}\} \cup \{\alpha = 0.0402\,\mathrm{M}^{-1}\}$ and its perturbations (in each perturbation, a parameter is multiplied by 1.05). In FPEI, we also use θ_0' to generate fixed paths. In another set of experiments (Figs. 4b, d, f, h, and 5b), we adapt parameter set $\theta_0'' = \{A_l = 468832.1058\,\mathrm{s}^{-1/2}, E_l = 2\,\mathrm{kcal\,mol}^{-1} \mid \forall l \in \mathcal{C}\} \cup \{\alpha = 0.0402\,\mathrm{M}^{-1}\}$ from θ_0' to increase the initial MSE in all experiments. We initialize the simplex in the Nelder-Mead algorithm with θ_0'' and its perturbations (in each perturbation, a parameter is multiplied by 1.05). In FPEI, we also generate fixed paths with θ_0''. In SSAI, we run the optimization until a limit on the number of iterations is reached or until a time limit is reached, which ever comes first. We also use this as the first stopping criteria in FPEI. In FPEI, to reduce the bias and to ensure that the ensemble of paths is a fair sample with respect to the optimized parameters, occasionally, the fixed paths are rebuilt from scratch and the optimization restarts. To this end, we set the second stopping criteria in FPEI to 200 iterations or 200 function evaluations of the Nelder-Mead algorithm, whichever comes first. Note that this empirical value is subject to change for different experiments. We could improve the method, by investigating a more robust way of when to update the paths. For example, we could compare the performance of SSA with the fixed paths in shorter intervals and update the fixed paths when their predictive quality diverges from SSA. During the optimization, we use an infinite value for parameter sets that have rates that are too slow or too fast; we bound downhill unimolecular and bimolecular rates (Eq. (7) and Eq. (8) of [42]) in $[1 \times 10^4, 1 \times 10^9]$ s^{-1} and in $[1 \times 10^4, 1 \times 10^9]\,\mathrm{M}^{-1}\mathrm{s}^{-1}$, respectively.

In Figs. 4d–h, FPEI reaches a minimal MSE more quickly than SSAI; consider the average computation time per iteration multiplied by the number of iterations to reach a minimal MSE. However, in Figs. 4a–c, SSAI reaches a minimal MSE more quickly than FPEI. This is because in Figs. 4d–h, the number of unique states is significantly smaller than the length of the paths. For example, in Fig. 4h, in the first set of fixed paths, the average length of a path is more than 2.3×10^7, whereas the average number of unique states and transitions is less than 1.5×10^5 and 5.6×10^5, respectively. In Fig. 4a, the average length of a path is 4.6×10^2, whereas the average number of unique states and transitions is 1.3×10^2 and 2.4×10^2, respectively. In Figs. 4e and f, which are slow dissociation reactions, compared to SSAI, FPEI speeds up parameter estimation by

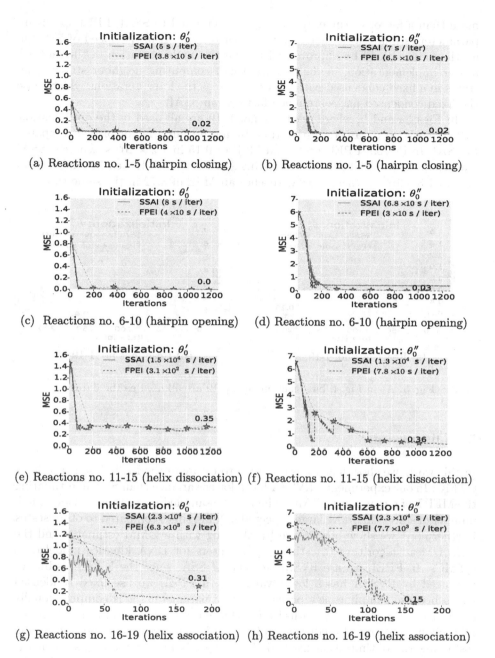

(a) Reactions no. 1-5 (hairpin closing)   (b) Reactions no. 1-5 (hairpin closing)

(c) Reactions no. 6-10 (hairpin opening)   (d) Reactions no. 6-10 (hairpin opening)

(e) Reactions no. 11-15 (helix dissociation) (f) Reactions no. 11-15 (helix dissociation)

(g) Reactions no. 16-19 (helix association) (h) Reactions no. 16-19 (helix association)

Fig. 4. The MSE of SSAI and FPEI on different types of reactions from Table 1. The average computation time per iteration is shown in the label of each method. The ★ markers show the MSE when trajectories are rebuilt from scratch using the learned parameter set from FPEI. In Figs. 4e–h, the SSAI traces stop at earlier iterations than the FPEI traces, even though SSAI was allocated more time than FPEI.

more than a factor of three. In Figs. 4g–h, compared to SSAI, FPEI speeds up parameter estimation by more than a factor of two. Also, the speed of FPEI in all the figures could be improved with a better implementation of the method; in our implementation, in the first iteration, computing neighbor states of all states in a fixed condensed path is slow, whereas the later iterations which reuse the fixed condensed paths are much faster than SSAI.

In Figs. 5a and b, where reactions no. 1–19 are all used in the optimization, FPEI speeds up parameter estimation, by more than a factor of two compared to SSAI. In Fig. 5a, FPEI reaches an MSE of 0.15 in 1.2×10^6 s, whereas SSAI reaches an MSE of 0.39 in the same time. In Fig. 5b, FPEI reaches an MSE of 0.43 in 1.3×10^6 s, whereas SSAI reaches an MSE of 3.72 in the same time.

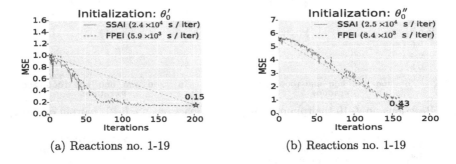

(a) Reactions no. 1-19   (b) Reactions no. 1-19

Fig. 5. As in Fig. 4, but reactions no. 1–19 are all used as the dataset.

5   Discussion

In this work, we show how to use RVSSA to estimate the MFPT of a reaction's CTMC. In our experiments, RVSSA has a lower variance than SSA in estimating the MFPT of a reaction's CTMC, when in the sampled paths there exists a small number of states that have large expected holding times compared to other states. Furthermore, we show how to use FPEI along with a GMM estimator and the Nelder-Mead algorithm to estimate parameters for DNA kinetics modeled as CTMCs. In FPEI, we use RVSSA instead of SSA, since the MFPT estimator produced by RVSSA has a lower variance. In FPEI, we use fixed condensed paths because sampling new paths for every parameter set is computationally expensive. Since using fixed paths leads to biased estimates, we alternate between minimizing the error of prediction on fixed paths, and resampling new paths and restarting the optimization method. FPEI speeds up computations when the number of unique states is significantly smaller than the length of sampled paths. In our experiments on a dataset of DNA reactions, FPEI speeds up parameter estimation compared to using SSAI, by more than a factor of three for slow reactions. Also, in our experiments, for reactions with large state spaces, it speeds up parameter estimation by more than a factor of two.

FPEI can be applied to reactions modeled as CTMCs, when the fixed paths can be produced in a timely manner. Generating paths for FPEI could be computationally expensive for slow reactions, such as helix dissociation from Morrison and Stols [25]. The runtime also depends on the parameter set that is used. It would be helpful to make FPEI applicable for such reactions, by speeding up the generation of the fixed paths, for example, with importance sampling approaches [13].

Finally, in this work, we evaluated FPEI in the context of DNA reactions. It would be useful to adopt and evaluate FPEI in other CTMC models [9,34,35], and other domains that require estimating MFPTs in CTMCs, such as protein folding.

References

1. Andrieu, C., Roberts, G.O.: The pseudo-marginal approach for efficient Monte Carlo computations. Ann. Stat. **37**, 697–725 (2009)
2. Andronescu, M., Aguirre-Hernandez, R., Condon, A., Hoos, H.H.: RNAsoft: a suite of RNA secondary structure prediction and design software tools. Nucleic Acids Res. **31**, 3416–3422 (2003)
3. Andronescu, M., Condon, A., Hoos, H.H., Mathews, D.H., Murphy, K.P.: Computational approaches for RNA energy parameter estimation. RNA **16**(12), 2304–2318 (2010)
4. Barton, R.R., Ivey Jr., J.S.: Nelder-Mead simplex modifications for simulation optimization. Manag. Sci. **42**(7), 954–973 (1996)
5. Bonnet, G., Krichevsky, O., Libchaber, A.: Kinetics of conformational fluctuations in DNA hairpin-loops. Proc. Natl. Acad. Sci. **95**(15), 8602–8606 (1998)
6. Cherry, K.M., Qian, L.: Scaling up molecular pattern recognition with DNA-based winner-take-all neural networks. Nature **559**(7714), 370 (2018)
7. Cisse, I.I., Kim, H., Ha, T.: A rule of seven in Watson-Crick base-pairing of mismatched sequences. Nat. Struct. Mol. Biol. **19**(6), 623 (2012)
8. Doucet, A., Johansen, A.M.: A tutorial on particle filtering and smoothing: fifteen years later. Handb. Nonlinear Filter. **12**(656–704), 3 (2009)
9. Flamm, C., Fontana, W., Hofacker, I.L., Schuster, P.: RNA folding at elementary step resolution. RNA **6**, 325–338 (2000)
10. Georgoulas, A., Hillston, J., Sanguinetti, G.: Unbiased Bayesian inference for population Markov jump processes via random truncations. Stat. Comput. **27**(4), 991–1002 (2017)
11. Gillespie, D.T.: Exact stochastic simulation of coupled chemical reactions. J. Phys. Chem. **81**(25), 2340–2361 (1977)
12. Golightly, A., Wilkinson, D.J.: Bayesian parameter inference for stochastic biochemical network models using particle Markov chain Monte Carlo. Interface Focus **1**(6), 807–820 (2011)
13. Hajiaghayi, M., Kirkpatrick, B., Wang, L., Bouchard-Côté, A.: Efficient continuous-time Markov chain estimation. In: International Conference on Machine Learning, pp. 638–646 (2014)
14. Hansen, L.P.: Large sample properties of generalized method of moments estimators. Econ. J. Econ. Soc. **50**, 1029–1054 (1982)
15. Hansen, L.P., Heaton, J., Yaron, A.: Finite-sample properties of some alternative GMM estimators. J. Bus. Econ. Stat. **14**(3), 262–280 (1996)

16. Hata, H., Kitajima, T., Suyama, A.: Influence of thermodynamically unfavorable secondary structures on DNA hybridization kinetics. Nucleic Acids Res. **46**(2), 782–791 (2017)
17. Hofacker, I.L.: Vienna RNA secondary structure server. Nucleic Acids Res. **31**(13), 3429–3431 (2003)
18. Hordijk, A., Iglehart, D.L., Schassberger, R.: Discrete time methods for simulating continuous time Markov chains. Adv. Appl. Probab. **8**(4), 772–788 (1976)
19. Horváth, A., Manini, D.: Parameter estimation of kinetic rates in stochastic reaction networks by the EM method. In: 2008 International Conference on BioMedical Engineering and Informatics, vol. 1, pp. 713–717. IEEE (2008)
20. Lehmann, E.L., Casella, G.: Theory of Point Estimation. Springer, New York (2006)
21. Loskot, P., Atitey, K., Mihaylova, L.: Comprehensive review of models and methods for inferences in bio-chemical reaction networks. arXiv preprint arXiv:1902.05828 (2019)
22. Lück, A., Wolf, V.: Generalized method of moments for estimating parameters of stochastic reaction networks. BMC Syst. Biol. **10**(1), 98 (2016)
23. Mathews, D.H., Sabina, J., Zuker, M., Turner, D.H.: Expanded sequence dependence of thermodynamic parameters improves prediction of RNA secondary structure. J. Mol. Biol. **288**(5), 911–940 (1999)
24. Metropolis, N., Rosenbluth, A.W., Rosenbluth, M.N., Teller, A.H., Teller, E.: Equation of state calculations by fast computing machines. J. Chem. Phys. **21**(6), 1087–1092 (1953)
25. Morrison, L.E., Stols, L.M.: Sensitive fluorescence-based thermodynamic and kinetic measurements of DNA hybridization in solution. Biochemistry **32**, 3095–3104 (1993)
26. Nelder, J.A., Mead, R.: A simplex method for function minimization. Comput. J. **7**(4), 308–313 (1965)
27. Ouldridge, T.E., Louis, A.A., Doye, J.P.: Structural, mechanical, and thermodynamic properties of a coarse-grained DNA model. J. Chem. Phys. **134**(8), 02B627 (2011)
28. Schaeffer, J.M.: Stochastic simulation of the kinetics of multiple interacting nucleic acid strands. Ph.D. thesis, California Institute of Technology (2012)
29. Schaeffer, J.M., Thachuk, C., Winfree, E.: Stochastic simulation of the kinetics of multiple interacting nucleic acid strands. In: Phillips, A., Yin, P. (eds.) DNA 2015. LNCS, vol. 9211, pp. 194–211. Springer, Cham (2015). https://doi.org/10.1007/978-3-319-21999-8_13
30. Schnoerr, D., Sanguinetti, G., Grima, R.: Approximation and inference methods for stochastic biochemical kinetics–a tutorial review. J. Phys. Math. Theor. **50**(9), 093001 (2017)
31. Schreck, J.S., et al.: DNA hairpins destabilize duplexes primarily by promoting melting rather than by inhibiting hybridization. Nucleic Acids Res. **43**(13), 6181–6190 (2015)
32. Srinivas, N., et al.: On the biophysics and kinetics of toehold-mediated DNA strand displacement. Nucleic Acids Res. **41**, 10641–10658 (2013)
33. Suhov, Y., Kelbert, M.: Probability and Statistics by Example: Volume 2, Markov Chains: A Primer in Random Processes and Their Applications, vol. 2. Cambridge University Press, Cambridge (2008)
34. Šulc, P., Romano, F., Ouldridge, T.E., Rovigatti, L., Doye, J.P., Louis, A.A.: Sequence-dependent thermodynamics of a coarse-grained DNA model. J. Chem. Phys. **137**(13), 135101 (2012)

35. Tang, X., Kirkpatrick, B., Thomas, S., Song, G., Amato, N.M.: Using motion planning to study RNA folding kinetics. J. Comput. Biol. **12**(6), 862–881 (2005)
36. Wackerly, D., Mendenhall, W., Scheaffer, R.L.: Mathematical Statistics with Applications. Cengage Learning, Boston (2014)
37. Wetmur, J.G.: Hybridization and renaturation kinetics of nucleic acids. Annu. Rev. Biophys. Bioeng. **5**(1), 337–361 (1976)
38. Xu, Z.Z., Mathews, D.H.: Experiment-assisted secondary structure prediction with RNAstructure. In: Turner, D., Mathews, D. (eds.) RNA Structure Determination: Methods and Protocols, pp. 163–176. Humana Press, New York (2016)
39. Zadeh, J.N., et al.: NUPACK: analysis and design of nucleic acid systems. J. Comput. Chem. **32**, 170–173 (2011)
40. Zhang, D.Y., Winfree, E.: Control of DNA strand displacement kinetics using toehold exchange. J. Am. Chem. Soc. **131**, 17303–17314 (2009)
41. Zhang, J.X., et al.: Predicting DNA hybridization kinetics from sequence. Nat. Chem. **10**(1), 91 (2018)
42. Zolaktaf, S., et al.: Inferring parameters for an elementary step model of DNA structure kinetics with locally context-dependent arrhenius rates. In: Brijder, R., Qian, L. (eds.) DNA 2017. LNCS, vol. 10467, pp. 172–187. Springer, Cham (2017). https://doi.org/10.1007/978-3-319-66799-7_12

New Bounds on the Tile Complexity of Thin Rectangles at Temperature-1

David Furcy, Scott M. Summers$^{(\boxtimes)}$, and Christian Wendlandt

Computer Science Department,
University of Wisconsin Oshkosh, Oshkosh, WI 54901, USA
{furcyd,summerss,wendlc69}@uwosh.edu

Abstract. In this paper, we study the minimum number of unique tile types required for the self-assembly of thin rectangles in Winfree's abstract Tile Assembly Model (aTAM), restricted to temperature-1. Using Catalan numbers, planar self-assembly and a restricted version of the Window Movie Lemma, we derive a new lower bound on the tile complexity of thin rectangles at temperature-1 in 2D. Then, we give the first known upper bound on the tile complexity of "just-barely" 3D thin rectangles at temperature-1, where tiles are allowed to be placed at most one step into the third dimension. Our construction, which produces a unique terminal assembly, implements a just-barely 3D, zig-zag counter, whose base depends on the dimensions of the target rectangle, and whose digits are encoded geometrically, vertically-oriented and in binary.

1 Introduction

Intuitively, self-assembly is the process through which simple, unorganized components spontaneously combine, according to local interaction rules, to form some kind of organized final structure.

While nature exhibits numerous examples of self-assembly, researchers have been investigating the extent to which the power of nano-scale self-assembly can be harnessed for the systematic nano-fabrication of atomically-precise computational, biomedical and mechanical devices. For example, in the early 1980s, Seeman [15] exhibited an experimental technique for controlling nano-scale self-assembly known as "DNA tile self-assembly".

Erik Winfree's abstract Tile Assembly Model (aTAM) is a simple, discrete mathematical model of DNA tile self-assembly. In the aTAM, a DNA tile is represented as an un-rotatable unit square *tile*. Each side of a tile may have a *glue* that consists of an integer strength, usually 0, 1 or 2, and an alphanumeric label. The idea is that, if two tiles abut with matching kinds of glues, then they bind with the strength of the glue. In the aTAM, a tile set may consist of a finite number of tiles, because individual DNA tiles are expensive to manufacture. However, an infinite number of copies of each tile are assumed to be available during the self-assembly process in the aTAM. Self-assembly starts by designating a *seed* tile and placing it at the origin. Then, a tile can come

© Springer Nature Switzerland AG 2019
C. Thachuk and Y. Liu (Eds.): DNA 25, LNCS 11648, pp. 100–119, 2019.
https://doi.org/10.1007/978-3-030-26807-7_6

in and bind to the seed-containing *assembly* if it binds with total strength at least a certain experimenter-chosen, integer value called the *temperature* that is usually 1 or 2. Self-assembly proceeds as tiles come in and bind one-at-a-time in an asynchronous and non-deterministic fashion.

Tile sets are designed to work at a certain temperature value. For instance, a tile set that self-assembles correctly at temperature-2 will probably not self-assemble correctly at temperature-1. However, if a tile set works correctly at temperature-1, then it can be easily modified to work correctly at temperature-2 (or higher). In what follows, we will refer to a "temperature-2" tile set as a tile set that self-assembles correctly only if the temperature is 2 and a "temperature-1" tile set as a tile set that self-assembles correctly if the temperature is 1.

Temperature-2 tile sets give the tile set designer more control over the order in which tiles bind to the seed-containing assembly. For example, in a temperature-2 tile set, unlike in a temperature-1 tile set, the placement of a tile at a certain location can be prevented until after the placements of at least two other tiles at two respective adjacent locations. This is known as *cooperative binding*. Cooperative binding in temperature-2 tile sets leads to the self-assembly of computationally and geometrically interesting structures, in the sense of Turing universality [17], the efficient self-assembly of $N \times N$ squares [1,12] and algorithmically-specified shapes [16].

While it is not known whether the results cited in the previous paragraph hold for temperature-1 tile sets, the general problem of characterizing the power of non-cooperative tile self-assembly is important from both a theoretical and practical standpoint. This is because when cooperative self-assembly is implemented in the laboratory [3,9,13,14,18], erroneous non-cooperative binding events may occur, leading to the production of invalid final structures. Of course, the obvious way to minimize such erroneous non-cooperative binding events is for experimenters to always implement systems that work in non-cooperative self-assembly because temperature-1 tile sets will work at temperature-1 or temperature-2. Yet, how capable is non-cooperative self-assembly, in general, or even in certain cases? At the time of this writing, no such general characterization of the power of non-cooperative self-assembly exists, but there are numerous results that show the apparent weakness of specific classes of temperature-1 tile self-assembly [5,8,10,11].

Although these results highlight the weakness of certain types of temperature-1 tile self-assembly, if 3D (unit cube) tiles are allowed to be placed in just-barely-three-dimensional Cartesian space (where tiles may be placed in just the $z = 0$ and $z = 1$ planes), then temperature-1 self-assembly is nearly as powerful as its two-dimensional cooperative counterpart. For example, like 2D temperature-2 tile self-assembly, just-barely 3D temperature-1 tile self-assembly is capable of simulating Turing machines [4] and the efficient self-assembly of squares [4,6] and algorithmically-specified shapes [7].

Furthermore, Aggarwal, Cheng, Goldwasser, Kao, Moisset de Espanés and Schweller [2] studied the efficient self-assembly of $k \times N$ rectangles, where $k < \frac{\log N}{\log \log N - \log \log \log N}$ at temperature-2 in 2D (the upper bound on k makes such a rectangle *thin*). They proved that the size of the smallest set of tiles that uniquely self-assemble into (i.e., the *tile complexity* of) a thin $k \times N$ rectangle is $O\left(N^{\frac{1}{k}} + k\right)$ and $\Omega\left(\frac{N^{\frac{1}{k}}}{k}\right)$ at temperature-2. Their lower bound actually applies to all tile sets (temperature-1, temperature-2, etc.) but their upper bound construction requires temperature-2 and does not work correctly at temperature-1.

In this paper, we continue the line of research into the tile complexity of thin rectangles, initiated by Aggarwal, Cheng, Goldwasser, Kao, Moisset de Espanés and Schweller, but exclusively for temperature-1 tile self-assembly.

1.1   Main Results of This Paper

The main results of this paper are bounds on the tile complexity of thin rectangles at temperature-1. We give an improved lower bound for the tile complexity of 2D thin rectangles as well as a non-trivial upper bound for the tile complexity of just-barely 3D thin rectangles. Intuitively, a just-barely 3D thin rectangle is like having at most two 2D thin rectangles stacked up one on top of the other. We prove two main results: one negative (lower bound) and one positive (upper bound). Our main negative result gives a new and improved asymptotic lower bound on the tile complexity of a 2D thin rectangle at temperature-1, without assuming unique production of the terminal assembly (unique self-assembly).

Theorem 1. *The tile complexity of a $k \times N$ rectangle for temperature-1 tile sets is $\Omega\left(N^{\frac{1}{k}}\right)$.*

Currently, the best upper bound for the tile complexity of a $k \times N$ rectangle for temperature-1 tile sets is $N + k - 1$, and is obtained via a straightforward generalization of the "Comb construction" by Rothemund and Winfree (see Fig. 2b of [12]). So, while Theorem 1 currently does not give a tight bound, its proof technique showcases a novel application of Catalan numbers to proving lower bounds for temperature-1 self-assembly in 2D, and could be of independent interest.

Our main positive result is the first non-trivial upper bound on the tile complexity of just-barely 3D thin rectangles.

Theorem 2. *The tile complexity of a just-barely 3D $k \times N$ thin rectangle for temperature-1 tile sets is $O\left(N^{\frac{1}{\lfloor \frac{k}{3} \rfloor}} + \log N\right)$. Moreover, our construction produces a tile set that self-assembles into a unique final assembly.*

We say that our main positive result is the first non-trivial upper bound because a straightforward generalization of the aforementioned Comb construction would give an upper bound of $O(N + k)$ on the tile complexity of a just-barely 3D $k \times N$ thin rectangle.

1.2   Comparison with Related Work

Aggarwal, Cheng, Goldwasser, Kao, Moisset de Espanés and Schweller [2] give a general lower bound of $\Omega\left(\frac{N^{\frac{1}{k}}}{k}\right)$ for the tile complexity of a 2D $k \times N$ rectangle for temperature-τ tile sets. Our main negative result, Theorem 1, is an asymptotic improvement of this result for the special case of temperature-1 self-assembly.

Aggarwal, Cheng, Goldwasser, Kao, Moisset de Espanés and Schweller [2] also prove that the tile complexity of a 2D $k \times N$ thin rectangle for general positive temperature tile sets is $O\left(N^{\frac{1}{k}} + k\right)$. Our main positive result, Theorem 2, is inspired by but requires a substantially different proof technique from theirs. Our construction, like theirs, uses a just-barely 3D counter, the base of which depends on the dimensions of the target rectangle, but unlike theirs, ours self-assembles in a zig-zag manner and the digits of the counter are encoded geometrically, vertically-oriented and in binary.

2   Preliminaries

In this section, we briefly sketch a 3D version of Winfree's abstract Tile Assembly Model. Going forward, all logarithms in this paper are base-2.

Fix an alphabet Σ. Σ^* is the set of finite strings over Σ. Let $\mathbb{Z}$, $\mathbb{Z}^+$, and $\mathbb{N}$ denote the set of integers, positive integers, and non-negative integers, respectively. Let $d \in \{2, 3\}$.

A *grid graph* is an undirected graph $G = (V, E)$, where $V \subset \mathbb{Z}^d$, such that, for all $\{a, b\} \in E$, $a - b$ is a d-dimensional unit vector. The *full grid graph* of V is the undirected graph $G_V^f = (V, E)$, such that, for all $x, y \in V$, $\{x, y\} \in E \iff \|x - y\| = 1$, i.e., if and only if x and y are adjacent in the d-dimensional integer Cartesian space.

A d-dimensional *tile type* is a tuple $t \in (\Sigma^* \times \mathbb{N})^{2d}$, e.g., a unit square (cube), with four (six) sides, listed in some standardized order, and each side having a *glue* $g \in \Sigma^* \times \mathbb{N}$ consisting of a finite string *label* and a non-negative integer *strength*. We call a d-dimensional tile type merely a *tile type* when d is clear from the context.

We assume a finite set of tile types, but an infinite number of copies of each tile type, each copy referred to as a *tile*. A *tile set* is a set of tile types and is usually denoted as T.

A *configuration* is a (possibly empty) arrangement of tiles on the integer lattice $\mathbb{Z}^d$, i.e., a partial function $\alpha : \mathbb{Z}^d \dashrightarrow T$. Two adjacent tiles in a configuration *bind*, *interact*, or are *attached*, if the glues on their abutting sides are equal (in both label and strength) and have positive strength. Each configuration α induces a *binding graph* G_α^b, a grid graph whose vertices are positions occupied by tiles, according to α, with an edge between two vertices if the tiles at those vertices bind. For two non-overlapping configurations α and β, $\alpha \cup \beta$ is defined as the unique configuration γ satisfying, for all $x \in \operatorname{dom} \alpha$,

$\gamma(\boldsymbol{x}) = \alpha(\boldsymbol{x})$, for all $\boldsymbol{x} \in \text{dom } \beta$, $\gamma(\boldsymbol{x}) = \beta(\boldsymbol{x})$, and $\gamma(\boldsymbol{x})$ is undefined at any point $\boldsymbol{x} \in \mathbb{Z}^d \backslash (\text{dom } \alpha \cup \text{dom } \beta)$.

An *assembly* is a connected, non-empty configuration, i.e., a partial function $\alpha : \mathbb{Z}^d \dashrightarrow T$ such that $G^{\text{b}}_{\text{dom } \alpha}$ is connected and dom $\alpha \neq \emptyset$. Given $\tau \in \mathbb{Z}^+$, α is τ-*stable* if every cut-set of G^{b}_{α} has weight at least τ, where the weight of an edge is the strength of the glue it represents.[1] When τ is clear from context, we say α is *stable*. Given two assemblies α, β, we say α is a *subassembly* of β, and we write $\alpha \sqsubseteq \beta$, if dom $\alpha \subseteq \text{dom } \beta$ and, for all points $\boldsymbol{p} \in \text{dom } \alpha$, $\alpha(\boldsymbol{p}) = \beta(\boldsymbol{p})$.

A d-dimensional *tile assembly system* (TAS) is a triple $\mathcal{T} = (T, \sigma, \tau)$, where T is a tile set, $\sigma : \mathbb{Z}^d \dashrightarrow T$ is the finite, τ-stable, *seed assembly*, and $\tau \in \mathbb{Z}^+$ is the *temperature*.

Given two τ-stable assemblies α, β, we write $\alpha \to^{\mathcal{T}}_1 \beta$ if $\alpha \sqsubseteq \beta$ and $|\text{dom } \beta \backslash \text{dom } \alpha| = 1$. In this case we say α $\mathcal{T}$-*produces* β *in one step*. If $\alpha \to^{\mathcal{T}}_1 \beta$, dom $\beta \backslash \text{dom } \alpha = \{\boldsymbol{p}\}$, and $t = \beta(\boldsymbol{p})$, we write $\beta = \alpha + (\boldsymbol{p} \mapsto t)$. The $\mathcal{T}$-*frontier* of α is the set $\partial^{\mathcal{T}} \alpha = \bigcup_{\alpha \to^{\mathcal{T}}_1 \beta} (\text{dom } \beta \backslash \text{dom } \alpha)$, i.e., the set of empty locations at which a tile could stably attach to α. The t-*frontier* of α, denoted $\partial^{\mathcal{T}}_t \alpha$, is the subset of $\partial^{\mathcal{T}} \alpha$ defined as $\{ \boldsymbol{p} \in \partial^{\mathcal{T}} \alpha \mid \alpha \to^{\mathcal{T}}_1 \beta \text{ and } \beta(\boldsymbol{p}) = t \}$.

Let $\mathcal{A}^T$ denote the set of all assemblies of tiles from T, and let $\mathcal{A}^T_{<\infty}$ denote the set of finite assemblies of tiles from T. A sequence of $k \in \mathbb{Z}^+ \cup \{\infty\}$ assemblies $\boldsymbol{\alpha} = (\alpha_0, \alpha_1, \ldots)$ over $\mathcal{A}^T$ is a $\mathcal{T}$-*assembly sequence* if, for all $1 \leq i < k$, $\alpha_{i-1} \to^{\mathcal{T}}_1 \alpha_i$. The *result* of an assembly sequence $\boldsymbol{\alpha}$, denoted as $\text{res}(\boldsymbol{\alpha})$, is the unique limiting assembly (for a finite sequence, this is the final assembly in the sequence).

We write $\alpha \to^{\mathcal{T}} \beta$, and we say α $\mathcal{T}$-*produces* β (in 0 or more steps), if there is a $\mathcal{T}$-assembly sequence $\alpha_0, \alpha_1, \ldots$ of length $k = |\text{dom } \beta \backslash \text{dom } \alpha| + 1$ such that (1) $\alpha = \alpha_0$, (2) dom $\beta = \bigcup_{0 \leq i < k} \text{dom } \alpha_i$, and (3) for all $0 \leq i < k$, $\alpha_i \sqsubseteq \beta$. If k is finite then it is routine to verify that $\beta = \alpha_{k-1}$.

We say α is $\mathcal{T}$-*producible* if $\sigma \to^{\mathcal{T}} \alpha$, and we write $\mathcal{A}[\mathcal{T}]$ to denote the set of $\mathcal{T}$-producible assemblies.

An assembly α is $\mathcal{T}$-*terminal* if α is τ-stable and $\partial^{\mathcal{T}} \alpha = \emptyset$. We write $\mathcal{A}_\square[\mathcal{T}] \subseteq \mathcal{A}[\mathcal{T}]$ to denote the set of $\mathcal{T}$-producible, $\mathcal{T}$-terminal assemblies. If $|\mathcal{A}_\square[\mathcal{T}]| = 1$ then $\mathcal{T}$ is said to be *directed*.

In general, a d-dimensional shape is a set $X \subseteq \mathbb{Z}^d$. We say that a TAS $\mathcal{T}$ *self-assembles* X if, for all $\alpha \in \mathcal{A}_\square[\mathcal{T}]$, dom $\alpha = X$, i.e., if every terminal assembly produced by $\mathcal{T}$ places a tile on every point in X and does not place any tiles on points in $\mathbb{Z}^d \backslash X$. We say that a TAS $\mathcal{T}$ *uniquely self-assembles* X if $\mathcal{A}_\square[\mathcal{T}] = \{\alpha\}$ and dom $\alpha = X$.

In the spirit of [12], we define the *tile complexity* of a shape X at temperature τ, denoted by $K^\tau_{SA}(X)$, as the minimum number of distinct tile types of any TAS in which it self-assembles, i.e., $K^\tau_{SA}(X) = \min \{n \mid \mathcal{T} = (T, \sigma, \tau), |T| = n \text{ and } X \text{ self-assembles in } \mathcal{T}\}$. The *directed tile complexity* of a shape X at temperature τ, denoted by $K^\tau_{USA}(X)$, is the minimum number of distinct

---

[1] A *cut-set* is a subset of edges in a graph which, when removed from the graph, produces two or more disconnected subgraphs. The *weight* of a cut-set is the sum of the weights of all of the edges in the cut-set.

tile types of any TAS in which it uniquely self-assembles, i.e., $K_{USA}^{\tau}(X) = \min\{n \mid \mathcal{T} = (T, \sigma, \tau), |T| = n \text{ and } X \text{ uniquely self-assembles in } \mathcal{T}\}$.

3   Lower Bound

In this section, we prove Theorem 1, which is a lower bound on the tile complexity of 2D rectangles. So, going forward, let $k, N \in \mathbb{N}$. We say that $R_{k,N}^2$ is a 2D $k \times N$ *rectangle* if $R_{k,N}^2 = \{0, 1, \ldots, N-1\} \times \{0, 1, \ldots, k-1\}$. Throughout this section, we will denote $R_{k,N}^2$ as simply $R_{k,N}$. Our lower bound relies on the following observation regarding temperature-1 self-assembly.

Observation 1. *If $\mathcal{T} = (T, \sigma, 1)$ is a singly-seeded TAS in which some shape X self-assembles and $\alpha \in \mathcal{A}[\mathcal{T}]$ such that* dom $\alpha = X$, *then G_{α}^b contains a simple path s from the location of σ to any location of X and there is a corresponding (simple) assembly sequence $\vec{\alpha}$ that follows s by placing tiles on and only on locations in s.*

Since, in Observation 1, we do not necessarily assume that $\mathcal{T}$ uniquely produces X, there could be more than one assembly sequence for a given s. Throughout the rest of this section, unless stated otherwise, let $\mathcal{T} = (T, \sigma, 1)$ be a singly-seeded TAS in which $R_{k,N}$ self-assembles.

3.1   Window Movie Lemmas

To prove our lower bound, we will use a variation of the Window Movie Lemma (WML) by Meunier, Patitz, Summers, Theyssier, Winslow and Woods and a corollary thereof. In this subsection, we review standard notation [10] for and give the statements of the variation that we use in our lower bound proof.

A *window* w is a set of edges forming a cut-set of the full grid graph of $\mathbb{Z}^d$. Given a window w and an assembly α, a window that *intersects* α is a partitioning of α into two configurations (i.e., after being split into two parts, each part may or may not be disconnected). In this case we say that the window w cuts the assembly α into two configurations α_L and α_R, where $\alpha = \alpha_L \cup \alpha_R$. Given a window w, its translation by a vector Δ, written $w + \Delta$ is simply the translation of each one of w's elements (edges) by Δ.

For a window w and an assembly sequence $\vec{\alpha}$, we define a *glue window movie* M to be the order of placement, position and glue type for each glue that appears along the window w in $\vec{\alpha}$. Given an assembly sequence $\vec{\alpha}$ and a window w, the associated glue window movie is the maximal sequence $M_{\vec{\alpha},w} = (v_1, g_1), (v_2, g_2), \ldots$ of pairs of grid graph vertices v_i and glues g_i, given by the order of the appearance of the glues along window w in the assembly sequence $\vec{\alpha}$. Furthermore, if m glues appear along w at the same instant (this happens upon placement of a tile which has multiple sides touching w) then these m glues appear contiguously and are listed in lexicographical order of the unit vectors describing their orientation in $M_{\alpha,w}$. We use the notation $\mathcal{B}(M_{\alpha,w})$ to denote the *bond-forming submovie* of $M_{\alpha,w}$, which consists of only those steps of $M_{\alpha,w}$ that place glues that eventually form positive-strength bonds in the assembly res(α). We write $M_{\alpha,w} + \Delta$ to denote the translation of the glue window movie, that is $(v_1 + \Delta, g_1), (v_2 + \Delta, g_2), \ldots$.

Let w be a window that partitions α into two configurations α_L and α_R, and assume $w + \Delta$ partitions β into two configurations β_L and β_R. Assume that α_L, β_L are the sub-configurations of α and β containing the seed tile of α and β, respectively.

Lemma 1 (Standard Window Movie Lemma). *If* $\mathcal{B}(M_{\alpha,w}) = \mathcal{B}(M_{\beta,w+\Delta}) - \Delta$, *then the following two assemblies are producible:* $\alpha_L(\beta_R - \Delta) = \alpha_L \cup (\beta_R - \Delta)$ *and* $\beta_L(\alpha_R + \Delta) = \beta_L \cup (\alpha_R + \Delta)$.

By Observation 1, there exist simple assembly sequences $\vec{\alpha} = (\alpha_i \mid 0 \leq i < l)$ and $\vec{\beta} = (\beta_i \mid 0 \leq i < m)$, with $l, m \in \mathbb{Z}^+$ that place tiles along simple paths s and s', respectively, leading to results α and β, respectively. The notation $M_{\alpha,w} \upharpoonright s$ represents the *restricted* glue window submovie (*restricted to s*), which consists of only those steps of M that place glues that eventually form positive-strength bonds along s.

Corollary 1 (Restricted Window Movie Lemma). *If* $M_{\alpha,w} \upharpoonright s = (M_{\beta,w+\Delta} \upharpoonright s') - \Delta$, *then the following two assemblies are producible:* $\alpha_L(\beta_R - \Delta) = \alpha_L \cup (\beta_R - \Delta)$ *and* $\beta_L(\alpha_R + \Delta) = \beta_L \cup (\alpha_R + \Delta)$.

The proof of Corollary 1 is identical to that of Lemma 1 and therefore is omitted. The proof is identical because $\vec{\alpha}$ ($\vec{\beta}$) follows s (s') and $\tau = 1$, so glues that do not follow s (s') are not necessary for the self-assembly of α (β). Note that Lemma 1 and Corollary 1 both generalize to 3D. See Fig. 1 for examples of $M_{\alpha,w}$, $\mathcal{B}(M_{\alpha,w})$ and $M_{\alpha,w} \upharpoonright s$. We now turn our attention to counting the number of restricted glue window submovies.

3.2   Counting Procedure for Undirected Self-assembly in 2D

In this subsection, we develop a counting procedure that we will use to obtain an upper bound on the number of distinct restricted glue window submovies.

Let $\rho \in \mathcal{A}_\square[\mathcal{T}]$. By Observation 1, there exists a simple path s in G_ρ^b from the location of σ to any location in an extreme (i.e., leftmost or rightmost) column of $R_{k,N}$. For the remainder of this section, unless stated otherwise, let α denote any (simple) assembly sequence that follows a simple path from the location of σ to some location in the furthest extreme column of $R_{k,N}$. So, if σ is in the left half of $R_{k,N}$, then s will go from σ to some location in the rightmost column of $R_{k,N}$. If σ is in the right half of $R_{k,N}$, then s will go from σ to some location in the leftmost column of $R_{k,N}$. If σ is in the middle column of $R_{k,N}$, then s can go to either the leftmost or rightmost column of $R_{k,N}$.

Index the columns of $R_{k,N}$ from 1 (left) to N (right). Assume c_σ is the index of the column in which the seed is contained. Consider two consecutive columns,

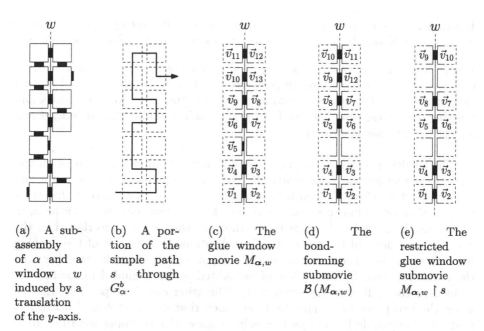

(a) A sub-assembly of α and a window w induced by a translation of the y-axis.

(b) A portion of the simple path s through G_α^b.

(c) The glue window movie $M_{\alpha,w}$.

(d) The bond-forming submovie $\mathcal{B}(M_{\alpha,w})$.

(e) The restricted glue window submovie $M_{\alpha,w} \upharpoonright s$.

Fig. 1. An assembly, a simple path and the various types of glue window movies.

with indices c_0 and c_1, satisfying $|c_\sigma - c_0| < |c_\sigma - c_1|$, and such that either c_0 or c_1 (or both) are in between c_σ and the furthest (from σ) extreme column. Since s is a simple path from the location of σ to some location in the extreme column of $R_{k,N}$ that is furthest from σ, s crosses between c_0 and c_1 through e *crossing edges* in G_ρ^b, where $1 \leq e \leq k$ and e is odd, visiting a total of $2e$ endpoints. The endpoint of a crossing edge in column c_0 (c_1) is its *near* (far) endpoint. A crossing edge points *away from* (towards) the seed if its near endpoint is visited first (second).

Observe that the first and last crossing edges visited by s must point away from the seed but each crossing edge that points away from the seed (except the last crossing edge) is immediately followed by a corresponding crossing edge that points towards the seed, when skipping the part of s that connects them without going through another crossing edge in between them. Assume that the rows of $R_{k,N}$ are assigned an index from 1 (top) to k (bottom). Let $E \subseteq \{1, \ldots, k\}$ be such that $|E| = e$ and $f \in E$ and $l \in E$ be the row indices of the first and last crossing edges visited, respectively. We define a *near* (far) *crossing pairing over E starting at $f \in E$* (ending at $l \in E$) as a set of $p = \frac{e-1}{2}$ non-overlapping pairs of elements in $E \setminus \{f\}$ ($E \setminus \{l\}$), where each pair contains (the row indices of) one crossing edge pointing away from the seed and its corresponding crossing edge pointing towards the seed. See Fig. 2 for examples of the previous definitions. For counting purposes pertaining to a forthcoming argument, we establish an injective mapping from near crossing pairings over E to strings of balanced parentheses of length $e - 1$.

Lemma 2. *There exists an injective function from the set of all near crossing pairings over E starting at f into the set of all strings of $2p$ balanced parentheses.*

Proof. Given a near crossing pairing P with p pairs, build a string x with $2p$ characters indexed from 1 to $2p$ going from left to right, as follows. For each element $\{a, b\}$ of P, with $a < b$ and assuming a is the i-th lowest row index and b is the j-th lowest row index in $E \backslash \{f\}$, place a left parenthesis at index i and a right parenthesis at index j.

The resulting string x contains exactly p pairs of parentheses. Furthermore, all of the parentheses in x are balanced because each opening parenthesis appears to the left of its closing parenthesis (thanks to the indexing used in the string construction algorithm just described) and, for any two pairs of parentheses in the string, it must be the case that either (1) they do not overlap (i.e., the closing parenthesis of the leftmost pair is positioned to the left of the opening parenthesis of the rightmost pair) or (2) one pair is nested inside the other (i.e., the interval defined by the indices of the nested pair is included in the interval defined by the indices of the outer pair). The other case is impossible, that is, when the two pairs $\{a, b\}$ and $\{c, d\}$ are such that $a < c < b < d$ because it would be impossible for any path crossing consecutive columns according to P to be simple. Consider any simple path $\pi_{\{a,b\}}$ that links crossing edges a and b without going through another crossing edge in between them. Since P is a near crossing pairing, $\pi_{\{a,b\}}$ is fully contained in the half-plane H_0 on the near side of c_0 toward the seed. This path partitions H_0 into two spaces. Since the crossing edges c and d belong to different components of this partition, any simple path linking these two crossing edges must cross $\pi_{\{a,b\}}$. Therefore, s, crossing between c_0 and c_1, could not have been simple.

Finally, note that two different near crossing pairings P_1 and P_2 over E starting at f will map to two different strings of balanced parentheses, so the mapping is injective.   □

Corollary 2. *There exists an injective function from the set of all far crossing pairings over E ending at l into the set of all strings of $2p$ balanced parentheses.*

The following is a systematic procedure for upper bounding the number of ways to select and order the $2e$ endpoints of the e crossing edges between an arbitrary pair of consecutive columns c_0 and c_1 (see Fig. 2a). Obviously, the number of ways to do this is less than or equal to $\binom{k}{e}(2e)!$. We will use crossing pairings and Catalan numbers to reduce this upper bound.

1. Choose the set E of row indices of the e crossing edges, out of k possible edges between consecutive columns. There are $\binom{k}{e}$ ways to do this.
2. One of the crossing edges must be first, so, choose the first crossing edge f. There are e ways to do this.
3. One of the crossing edges must be last, so, choose the last crossing edge l. There are e ways to do this.

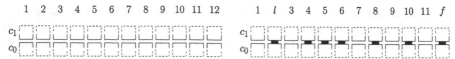

(a) Two consecutive columns, for "height" $k = 12$, rotated 90 degrees counterclockwise. We show the columns this way for ease of examining the corresponding strings of balanced parentheses.

(b) Steps 1, 2 and 3. Here, $f = 12$, $l = 2$ and $E = \{2, 4, 5, 6, 8, 10, 12\}$, with chosen glues indicated by the little black rectangles.

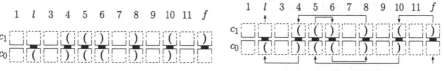

(c) Steps 4 and 5. Here, $x_0 = ()(())$ and $x_1 = (())()$.

(d) The π with $|\pi| = 14 = 2e$ that is found by the algorithm in Step 6.

Fig. 2. A "good" sample run of the counting procedure.

The three previous steps are depicted in Fig. 2b. We purposely allow choosing $l = f$ because our intention is to upper bound the number of ways to select and order the endpoints of the crossing edges. Moreover, $l = f$ when $e = 1$. Denote as $C_p = \frac{1}{p+1}\binom{2p}{p}$ the pth Catalan number (indexed starting at 0).

4. For a given pair of consecutive columns, in which f is visited first, s induces a near crossing pairing over E starting at f, where the elements of each pair are row indices of near endpoints of crossing edges, including l, but not f. By Lemma 2, it suffices to count the number of ways to choose a string of $2p$ balanced parentheses. Therefore, choose a string x_0 of $2p$ balanced parentheses, where $x_0[i]$ corresponds to the crossing edge in E with the i-th lowest row index, excluding f. There are C_p ways to do this.

5. For a given pair of consecutive columns in which l is visited last, s induces a far crossing pairing over E ending at l, where the elements of each pair are row indices of far endpoints of crossing edges, including f, but not l. By Corollary 2, it suffices to count the number of ways to choose a string of $2p$ balanced parentheses. Therefore, choose a string x_1 of $2p$ balanced parentheses, where $x_1[i]$ corresponds to the crossing edge in E with the i-th lowest row index, excluding l. There are C_p ways to do this.

The two previous steps are depicted in Fig. 2c. At this point, we have chosen the locations and connectivity pattern of all the crossing edges. We now show that there is at most one way in which both endpoints of every crossing edge may be visited by s, subject to the constraints imposed by the previous steps.

6. Let $I_j(r)$ be the index i, such that $x_j[i]$ corresponds to the crossing edge with row index r. The following greedy algorithm attempts to build a path

π of locations that (1) starts at the near endpoint of the first crossing edge, (2) ends at the far endpoint of the last crossing edge and (3) visits only the endpoints of the crossing edges while following the balanced parenthesis pairings of both x_0 and x_1.

1 Initialize $\pi = ((c_0, f), (c_1, f))$ to be a sequence of locations, $j = 1$ and $r = f$, where r stands for "row number" and j stands for the current side, near (0) or far (1).

2 **while** $r \neq l$ **do**

3 $\quad$ Let r' be the unique row index of the crossing edge, such that, $I_j(r)$ and $I_j(r')$ are paired.

4 $\quad$ Set $r = r'$.

5 $\quad$ Append (c_j, r) to π.

6 $\quad$ Set $j = (j + 1) \mod 2$.

7 $\quad$ Append (c_j, r) to π.

First, note that no endpoint can be visited more than once, so the algorithm always terminates. Second, note that, when the algorithm terminates, either $|\pi| = 2e$ (see Fig. 2d) or $|\pi| < 2e$ (see Fig. 3). However, regardless of its length, π is the unique sequence of endpoints of crossing edges that can be visited by any simple path starting at (c_0, f), ending at (c_1, l) and following the balanced parenthesis pairings of both x_0 and x_1. The uniqueness of π follows from the uniqueness of r', given r, based on the balanced parenthesis pairings of both x_0 and x_1.

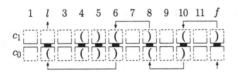

Fig. 3. A "bad" sample run of the counting procedure. Here, $x_0 = (())()$ and $x_1 = ()()()$. With $f = 12$, $l = 2$ and $E = \{2, 4, 5, 6, 8, 10, 12\}$, there is no valid π (crossing edges 4 and 5 are skipped, assuming the pairs of parentheses are faithfully followed), so the algorithm terminates with $|\pi| = 10 < 2e = 14$.

By the above counting procedure, there are at most $\binom{k}{e}\left(e\frac{1}{p+1}\binom{2p}{p}\right)^2 \cdot 1$ ways to select and order the endpoints of the crossing edges between an arbitrary pair of consecutive columns c_0 and c_1 as they are visited by a simple path.

3.3 Lower Bound for Undirected Self-assembly in 2D: Theorem 1

To prove a lower bound on $K_{SA}^1(R_{k,N})$, we turn our attention to upper bounding the number of restricted glue window submovies of the form $M_{\alpha,w} \restriction s$. For the remainder of this subsection, assume w is always some window induced by (a translation of) the y-axis that cuts $R_{k,N}$ between some pair of consecutive columns.

Lemma 3. *The number of restricted glue window submovies of the form* $M_{\alpha,w} \restriction s$ *is less than or equal to* $|G|^k \cdot 2^{3k+2} \cdot k$, *where* G *is the set of all glues of (the tile types in)* T.

Proof. Let e be an odd number such that $1 \leq e \leq k$ and $M_{\alpha,w} \restriction s = (v_1, g_1), \ldots, (v_{2e}, g_{2e})$ be a restricted glue window submovie. Since α follows a simple path, $g_{2i-1} = g_{2i}$ for $i = 1, \ldots, e$. This means that we only need to assign e glues, with $|G|$ choices for each glue. So, the number of ways to assign glues in $M_{\alpha,w} \restriction s$ is less than or equal to $|G|^e$. Since α follows a simple path, each location in $M_{\alpha,w} \restriction s$ corresponds to an endpoint of a crossing edge that crosses w. So, the number of ways to assign locations in $M_{\alpha,w} \restriction s$ is less than or equal to the number of ways to select and order the endpoints of e crossing edges that cross w via α. By the above counting procedure, the number of ways to select and order the endpoints of e crossing edges that cross w via α is less than or equal to $\binom{k}{e} \left(e \frac{1}{p+1} \binom{2p}{p} \right)^2$. Thus, if m is the total number of restricted glue window submovies of the form $M_{\alpha,w} \restriction s$, then we have:

$$
m \leq \sum_{\substack{1 \leq e \leq k \\ e \text{ odd}}} \left(\binom{k}{e} \left(e \frac{1}{p+1} \binom{2p}{p} \right)^2 |G|^e \right) \leq |G|^k \cdot 2^{3k+2} \cdot k
$$

$\square$

We will use Lemma 3 to prove our impossibility result.

Theorem 1. $K_{SA}^1 (R_{k,N}) = \Omega \left(N^{\frac{1}{k}} \right)$.

Proof. Let G be the set of glues of (the tile types in) T. It suffices to show that $|T| = \Omega \left(N^{\frac{1}{k}} \right)$. Since s is the longest path in G_p^b, from the location of σ to some location in an extreme column of $R_{k,N}$, by Lemma 3, if $N > 2 \cdot |G|^k \cdot 2^{3k+2} \cdot k$, then there exists a window w, along with vectors Δ_1 and Δ_2, such that, $\Delta_1 \neq 0$, $\Delta_2 \neq 0$ and $\Delta_1 \neq \Delta_2$ and satisfying $M_{\alpha,w} \restriction s = (M_{\alpha,w+\Delta_1} \restriction s) - \Delta_1$ and $M_{\alpha,w+\Delta_1} \restriction s = (M_{\alpha,w+\Delta_2} \restriction s) - \Delta_2$. Without loss of generality, assume that w and $w + \Delta_1$ are on the same side of σ and $w + \Delta_1$ is to the left of w. Assume that w partitions α into α_L and α_R and $w + \Delta_1$ partitions $\beta = \alpha$ into β_L and β_R. Then, by Corollary 1, $\beta_L (\alpha_R + \Delta_1) \in A[T]$. Since $\Delta_1 \neq 0$, $\text{dom} (\beta_L (\alpha_R + \Delta_1)) \setminus R_{k,N} \neq \emptyset$. In other words, T produces some assembly that places at least one tile at a location that is not an element of $R_{k,N}$. Therefore, it must hold that $N \leq 2 \cdot |G|^k \cdot 2^{3k+2} \cdot k$, which implies that $|G|^k \geq \frac{N}{2^{3k+3} \cdot k}$, and thus $|G| \geq \left(\frac{N}{2^{3k+3} \cdot k} \right)^{\frac{1}{k}} \geq \frac{N^{\frac{1}{k}}}{(2^{6k} \cdot 2^k)^{\frac{1}{k}}} = \frac{N^{\frac{1}{k}}}{128}$. Finally, note that $|T| \geq \frac{|G|}{4}$ and it follows that $|T| = \Omega \left(N^{\frac{1}{k}} \right)$. $\square$

Note that the main technique for the proof of Theorem 1 does not hold in 3D because Lemma 2 requires planarity.

4 Upper Bound

In this section, we give a construction for a singly-seeded TAS in which a sufficiently large just-barely 3D rectangle uniquely self-assembles. Going forward, we say that $R^3_{k,N} \subseteq \mathbb{Z}^3$ is a 3D $k \times N$ *rectangle* if $\{0, 1, \ldots, N-1\} \times \{0, 1, \ldots, k-1\} \times \{0\} \subseteq R^3_{k,N} \subseteq \{0, 1 \ldots, N-1\} \times \{0, 1 \ldots, k-1\} \times \{0, 1\}$. For the sake of clarity of presentation, we represent $R^3_{k,N}$ vertically.

Following standard presentation conventions for "just-barely" 3D tile self-assembly, we use big squares to represent tiles placed in the $z = 0$ plane and small squares to represent tiles placed in the $z = 1$ plane. A glue between a $z = 0$ tile and a $z = 1$ tile is denoted as a small black disk. Glues between $z = 0$ tiles are denoted as thick lines. Glues between $z = 1$ tiles are denoted as thin lines. The following is our main positive result.

Theorem 2. $K^1_{USA}\left(R^3_{k,N}\right) = O\left(N^{\frac{1}{\lfloor \frac{k}{3} \rfloor}} + \log N\right)$.

The general idea of our construction is inspired by, but substantially different from, a similar construction by Aggarwal, Cheng, Goldwasser, Kao, Moisset de Espanés and Schweller [2] for the self-assembly of two-dimensional $k \times N$ thin rectangles at temperature-2. The basic idea of our construction for Theorem 2 is to use a counter, the base of which is a function of k and N. Then, we initialize the counter with a certain starting value and have it increment until its maximum value, at which point it rolls over to all 0's and the assembly goes terminal. See Fig. 4 for an example of our construction shown at two different levels of granularity.

Since the height (number of tile rows) of each logical row in the counter depends on k and N, we must choose its starting value carefully. Therefore, let $d = \lfloor \frac{k}{3} \rfloor$, $m = \left\lceil \left(\frac{N}{5}\right)^{\frac{1}{d}} \right\rceil$, $l = \lceil \log m \rceil + 1$, $s = m^d - \left\lfloor \frac{N-3l-1}{3l+2} \right\rfloor$, $c = k \mod 3$, and $r = N + 1 \mod 3l + 2$, where d is the number of digits in the counter, m is the numerical base of the counter, l is the number of bits needed to represent each digit in the counter's value plus one for the "left edge", s is the numerical start of the counter, and c and r are the number of tile columns and tile rows, respectively, that must be filled in after and outside of the counter. Each digit of the counter requires a width of 3 tiles, which has a direct relation with the tile complexity of the construction. The values of m and s are chosen such that the counter stops just before reaching a height of N tiles, at which point, the assembly is given a flat "roof". For example, in Fig. 4, we have $d = 3$, $m = 3$, $l = 3$, $s = 23$, $c = 2$, and $r = 2$.

A *gadget* is a group of tiles that perform a specific task as they self-assemble. Gadgets are referred to by a name like Gadget. We define the tile set for our construction that proves Theorem 2 via a series of gadgets. All gadgets are depicted in a corresponding figure, where the input glue is explicitly specified by an arrow, output glues are inferred and glues internal to the gadget are configured to ensure unique self-assembly within the gadget. We say that a gadget is *general* if its input and output glues are undefined. From general gadgets, we create

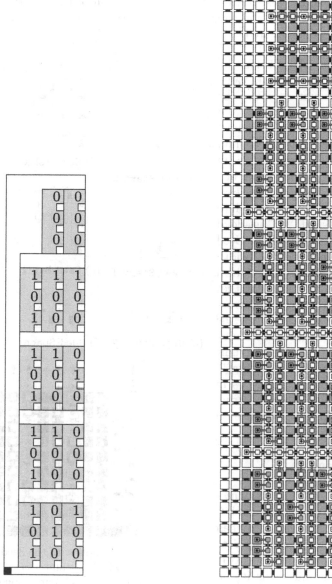

(a) High-level: values of the counter.   (b) Low-level: full example.

Fig. 4. A full example of a 11×56 construction. The counter begins at 10-01-10 and is counting in ternary, so the initial value is $23 = 2 \cdot 3^2 + 1 \cdot 3^1 + 2 \cdot 3^0$. Note that the least significant bit for each digit, which is the lowest bit in each digit column, is actually a "left edge" indicator, where "1" means "leftmost" and "0" means "not leftmost". To help distinguish overlapping tiles, "write gadgets" are drawn in gray.

specific gadgets in which all of the glues are defined. We define a *gadget unit* as a collection of gadgets with a singular purpose, like incrementing the value of a counter. In the rest of this section, we define all of the general gadgets, from which all of the specific gadgets that comprise all of the gadget units of our construction are created.

Seed Unit: We begin by encoding the initial value of the counter with the Seed unit. It has d 3-wide (logical/digit) columns, where each 3-wide column represents a digit of s in base-m. A collection of bit-bumps on the columns' east sides encodes the digits into binary. A small "lip" is added on the west side of the Seed unit to increase the width of the assembly by c, which catches any vertical filler tiles at the end of the construction. The Guess tile on the east side of the unit initiates the first Counter unit (defined next). See Fig. 5. The number of tile types in the Seed unit in our construction is $O(\log N)$ for general rectangles, or $O\left(N^{\frac{1}{\lfloor\frac{k}{3}\rfloor}}\right)$ for thin rectangles.

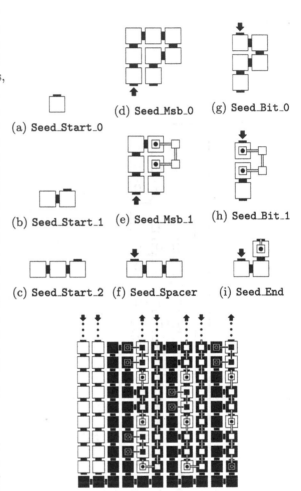

(a) Seed_Start_0

(b) Seed_Start_1

(c) Seed_Start_2

(d) Seed_Msb_0

(e) Seed_Msb_1

(f) Seed_Spacer

(g) Seed_Bit_0

(h) Seed_Bit_1

(i) Seed_End

(j) An example with the Seed unit colored black. The counter is set to 10-01-10, which is 23 in binary-coded-ternary. The seed is the leftmost tile in the bottom row and self-assembly of the black tiles proceeds in a left-to-right fashion.

Fig. 5. The Seed gadget unit. The actual seed tile is at the far-left of any of the Seed_Start gadgets.

Counter Unit: Each Counter unit reads over a series of bit-bumps protruding into their row from the preceding Seed unit or counter row. After a Counter unit reads its bit pattern with Guess tiles, the unit produces a new bit pattern in the row above it that encodes a copy or increment of the current digit. Of the "less significant digit" units, the one that increments $m-1$

to 0 is unique because it initiates another increment unit, that is, a carry is passed to the digit to its left. Other increment units, as well as the copy units, will only initiate copy units (no carry is propagated to the left). The first bit read is always the "left edge" marker, which tells the unit if it represents the most significant digit of the counter value and needs to start a new row instead of another Counter unit. The counter terminates when the most significant digit follows an increment unit and reads $m - 1$ in its column. At that point, the counter will have rolled over $m^d - 1$ and the Roof unit (defined later) takes over. The gadgets belonging to the Counter units are shown in Fig. 6. The number of

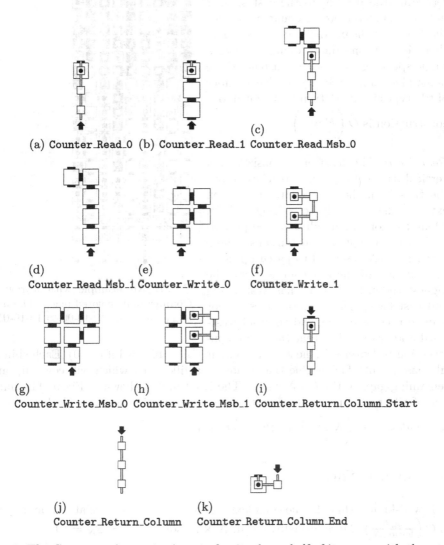

(a) Counter_Read_0 (b) Counter_Read_1 Counter_Read_Msb_0
                                                    (c)

(d)   (e)   (f)
Counter_Read_Msb_1 Counter_Write_0   Counter_Write_1

(g)   (h)   (i)
Counter_Write_Msb_0 Counter_Write_Msb_1 Counter_Return_Column_Start

(j)   (k)
Counter_Return_Column Counter_Return_Column_End

Fig. 6. The Counter gadget unit. A row of units shares half of its space with the row above and the other half of its space with the row below.

tile types in the Counter unit in our construction is $O\left(N^{\frac{1}{\lfloor \frac{k}{3} \rfloor}}\right)$ and see Fig. 7 for an example.

Return Row Unit: To begin a new row of Counter units following the completion of the current row, a `Return Row` gadget unit must return the frontier of the assembly to the east side of the construction. After returning the frontier to the east side, it initiates an incrementing `Counter` unit for the least significant digit. In cases where there is only one digit column in use, a single special gadget is used instead of the multi-piece unit. See Fig. 8. The number of tile types in the Return Row unit in our construction is $O\left(N^{\frac{1}{\lfloor \frac{k}{3} \rfloor}}\right)$.

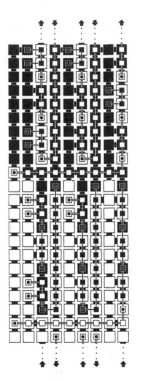

Fig. 7. An example with a row of Counter units colored black. The unit is incrementing 10-10-00 to 10-10-01.

Roof Unit: The `Roof` unit consists of a vertical tile column that reaches above the tiles from the last counter row, then extends the assembly to a height of N. Along the column are glues placed periodically that accept filler extensions. These extensions are designed to patch up holes that are left in the last counter row. The highest tile in the column has west-facing and east-facing glues. These glues accept shingle tiles, which extend the roof westward and eastward so that the entire construction is "covered" (the eastward expansion is blocked if $r = 0$). Each shingle tile has a south-facing glue that binds to a filler tile, which will cover up any remaining gaps in the $k \times N$ shape. The Roof unit is shown in Fig. 9. The number of tile types in the Roof unit in our construction is $O(\log N)$ for general rectangles, or $O\left(N^{\frac{1}{\lfloor \frac{k}{3} \rfloor}}\right)$ for thin rectangles.

5   Future Work

It is well-known that the tile complexity of a 2D $N \times N$ square at temperature-2 is $O\left(\frac{\log N}{\log \log N}\right)$. More formally, for an $N \times N$ square $S_N^2 = S_N = \{0, 1, \ldots, N-1\} \times \{0, 1, \ldots, N-1\}$, $K_{USA}^2(S_N) = O\left(\frac{\log N}{\log \log N}\right)$ [1]. However, it is conjectured [8] that $K_{USA}^1(S_N) = 2N - 1$, meaning that a 2D $N \times N$ square does

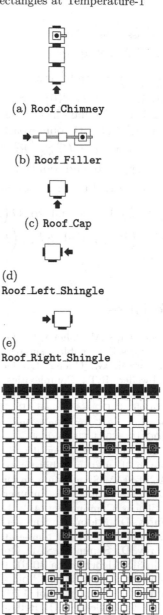

(a) Roof_Chimney

(b) Roof_Filler

(c) Roof_Cap

(d)
Roof_Left_Shingle

(e)
Roof_Right_Shingle

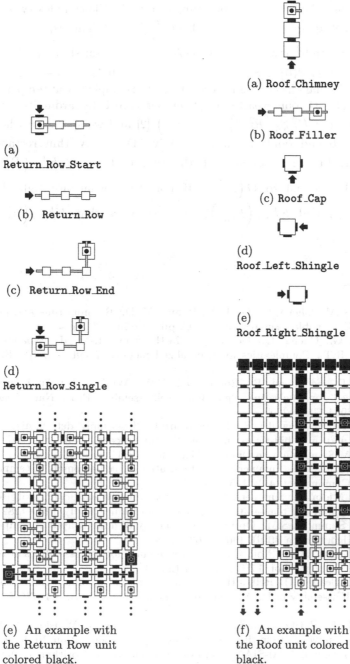

(a)
Return_Row_Start

(b) Return_Row

(c) Return_Row_End

(d)
Return_Row_Single

(e) An example with
the Return Row unit
colored black.

Fig. 8. The Return Row gadget unit.

(f) An example with
the Roof unit colored
black.

Fig. 9. The Roof gadget unit.

not self-assemble efficiently at temperature-1. Yet, the tile complexity of a just-barely 3D $N \times N$ square at temperature-1 is $O\left(\frac{\log N}{\log \log N}\right)$. That is, $K_{USA}^1\left(S_N^3\right) = O\left(\frac{\log N}{\log \log N}\right)$, where S_N^3 is a just-barely 3D $N \times N$ square, satisfying $\{0, 1, \ldots, N-1\} \times \{0, 1, \ldots, N-1\} \times \{0\} \subseteq S_N^3 \subseteq \{0, 1, \ldots, N-1\} \times \{0, 1, \ldots, N-1\} \times \{0, 1\}$ [6]. So, a 2D $N \times N$ square has the same asymptotic tile complexity at temperature-2 as its just-barely 3D counterpart does at temperature-1. Regarding thin rectangles, we know that $K_{USA}^2\left(R_{k,N}^2\right) = O\left(N^{\frac{1}{k}}\right)$ [2] and we speculate whether a similar upper bound holds for a just-barely 3D $k \times N$ thin rectangle at temperature-1. In other words, is it the case that either $K_{SA}^1\left(R_{k,N}^3\right)$ or $K_{USA}^1\left(R_{k,N}^3\right)$ is equal to $O\left(N^{\frac{1}{k}}\right)$? If not, then what are tight bounds for $K_{SA}^1\left(R_{k,N}^3\right)$ and $K_{USA}^1\left(R_{k,N}^3\right)$? We conjecture that $K_{USA}^1\left(R_{k,N}^3\right) = o\left(N^{\frac{1}{\lfloor\frac{k}{3}\rfloor}}\right)$.

References

1. Adleman, L.M., Cheng, Q., Goel, A., Huang, M.-D.: Running time and program size for self-assembled squares. In: STOC, pp. 740–748 (2001)
2. Aggarwal, G., Cheng, Q., Goldwasser, M.H., Kao, M.-Y., de Espanés, P.M., Schweller, R.T.: Complexities for generalized models of self-assembly. SIAM J. Comput. **34**, 1493–1515 (2005)
3. Barish, R.D., Schulman, R., Rothemund, P.W., Winfree, E.: An information-bearing seed for nucleating algorithmic self-assembly. Proc. Nat. Acad. Sci. **106**(15), 6054–6059 (2009)
4. Cook, M., Fu, Y., Schweller, R.: Temperature 1 self-assembly: deterministic assembly in 3D and probabilistic assembly in 2D. In: Proceedings of the 22nd Annual ACM-SIAM Symposium on Discrete Algorithms (2011)
5. Doty, D., Patitz, M.J., Summers, S.M.: Limitations of self-assembly at temperature 1. Theor. Comput. Sci. **412**, 145–158 (2011)
6. Furcy, D., Micka, S., Summers, S.M.: Optimal program-size complexity for self-assembled squares at temperature 1 in 3D. Algorithmica **77**(4), 1240–1282 (2017)
7. Furcy, D., Summers, S.M.: Optimal self-assembly of finite shapes at temperature 1 in 3D. Algorithmica **80**(6), 1909–1963 (2018)
8. Manuch, J., Stacho, L., Stoll, C.: Two lower bounds for self-assemblies at temperature 1. J. Comput. Biol. **17**(6), 841–852 (2010)
9. Mao, C., LaBean, T.H., Reif, J.H., Seeman, N.C.: Logical computation using algorithmic self-assembly of DNA triple-crossover molecules. Nature **407**(6803), 493–496 (2000)
10. Meunier, P.-E., Patitz, M.J., Summers, S.M., Theyssier, G., Winslow, A., Woods, D.: Intrinsic universality in tile self-assembly requires cooperation. In: Proceedings of the 25th Annual ACM-SIAM Symposium on Discrete Algorithms (SODA), pp. pp. 752–771 (2014)
11. Meunier, P.-É., Woods, D.: The non-cooperative tile assembly model is not intrinsically universal or capable of bounded turing machine simulation. In: Proceedings of the 49th Annual ACM SIGACT Symposium on Theory of Computing, STOC 2017, Montreal, 19–23 June 2017, pp. 328–341 (2017)

12. Rothemund, P.W.K., Winfree, E.: The program-size complexity of self-assembled squares (extended abstract). In: STOC 2000: Proceedings of the Thirty-Second Annual ACM Symposium on Theory of Computing, pp. 459–468 (2000)
13. Rothemund, P.W.K., Papadakis, N., Winfree, E.: Algorithmic self-assembly of DNA Sierpinski triangles. PLoS Biol. **2**(12), 2041–2053 (2004)
14. Schulman, R., Winfree, E.: Synthesis of crystals with a programmable kinetic barrier to nucleation. Proc. Nat. Acad. Sci. **104**(39), 15236–15241 (2007)
15. Seeman, N.C.: Nucleic-acid junctions and lattices. J. Theor. Biol. **99**, 237–247 (1982)
16. Soloveichik, D., Winfree, E.: Complexity of self-assembled shapes. SIAM J. Comput. **36**(6), 1544–1569 (2007)
17. Winfree, E.: Algorithmic self-assembly of DNA. Ph.D. thesis, California Institute of Technology, June 1998
18. Winfree, E., Liu, F., Wenzler, L.A., Seeman, N.C.: Design and self-assembly of two-dimensional DNA crystals. Nature **394**(6693), 539–544 (1998)

Non-cooperatively Assembling Large Structures

Pierre-Étienne Meunier[1] and Damien Regnault[2(✉)]

[1] Maynooth University, Maynooth, Ireland
`pierre-etienne.meunier@mu.ie`
[2] IBISC, Univ Évry, Université Paris-Saclay, 91025 Evry, France
`damien.regnault@univ-evry.fr`

Abstract. Algorithmic self-assembly is the study of the local, distributed, asynchronous algorithms ran by molecules to self-organise, in particular during crystal growth. The general *cooperative* model, also called *"temperature 2"*, uses *synchronisation* to simulate Turing machines, build shapes using the smallest possible amount of tile types, and other algorithmic tasks. However, in the non-cooperative ("temperature 1") model, the growth process is entirely asynchronous, and mostly relies on geometry. Even though the model looks like a generalisation of finite automata to two dimensions, its 3D generalisation is capable of performing arbitrary (Turing) computation [SODA 2011], and of universal simulations [SODA 2014], whereby a single 3D non-cooperative tileset can simulate the dynamics of all possible 3D non-cooperative systems, up to a constant scaling factor.

However, the original 2D non-cooperative model is not capable of universal simulations [STOC 2017], and the question of its computational power is still widely open and it is conjectured to be weaker than "temperature 2" or its 3D counterpart. Here, we show an unexpected result, namely that this model can reliably grow assemblies of diameter $\Theta(n \log n)$ with only n tile types, which is the first asymptotically efficient positive construction.

Keywords: Self-assembly · aTAM · Temperature 1

1 Introduction

Our ability to understand and control matter at the scale of molecules conjures a future where we can engineer our own materials, interact with biological networks to cure their malfunctions, and build molecular computers and nanoscale

Research supported by European Research Council (ERC) under the European Union's Horizon 2020 research and innovation programme (grant agreement No 772766, Active-DNA project), and Science Foundation Ireland (SFI) under Grant number 18/ERCS/5746.

factories. The field of molecular computing and molecular self-assembly studies the algorithms run by molecules to exchange information, to self-organise in space, to grow and replicate. Our goal is to build a theory of how they compute, and of how we can program them.

One of the most successful models of algorithmic self assembly is the *abstract tile assembly model*, imagined by Winfree [21]. In that model, we start with a single seed assembly and a finite number of tile types (with an infinite supply of each type), and attach tiles, one at a time, asynchronously and nondeterministically, to the assembly, based on a condition on their borders' colours.

This model has served to bootstrap the field of molecular computing, which has since produced an impressive number of experimental realisations, from DNA motors [23] to arbitrary hard-coded shapes at the nanoscale [17], and cargo-sorting robots [20].

On the theoretical side, the abstract tile assembly model has been used to explore different features of self-organisation in space, especially in an asynchronous fashion. In many variants of the model, tile assembly is capable of simulating Turing machines [1–3,6,8,10,14–16,19,21,22]. More surprisingly, in its original form, the model is *intrinsically universal* [4], meaning that there is a single "universal" tileset capable of simulating the behaviour of any other tileset (that behaviour is encoded only in the seed of the universal tileset).

In the usual form of the model, a part of the assembly must "wait" for another to grow long enough to cooperate. In the non-cooperative model, however, any tile can attach to any location, as long as at least one side matches the colour of that location. Therefore, "synchronising" different parts of the assembly is impossible, and the main question becomes, *what kind of computation can we do in a completely asynchronous way?* The answer seems to depend crucially on the space in which the assemblies grow: in one dimension, non-cooperative tile assembly is equivalent to finite automata[1], and are therefore not too powerful. In three dimensions though, this model is capable of simulating Turing machines [2], and even of simulating itself *intrinsically* [12]. If instead of square tiles, we use tiles that do *not* tile the plane, the situation becomes even more puzzling: tiles whose shape are regular polygons can perform arbitrary computation, but only if they have at least seven sides [9]. In a similar way, polyomino tiles can also simulate Turing machines, provided that at least one of their dimensions is at least two [7].

However, in two dimensions with standard square tiles, the capabilities of this model remain largely mysterious. All we know is that it cannot simulate the general (cooperative) model up to rescaling [12], and cannot simulate itself either [13], but we know very little about its actual computational power. A number of related questions and simpler models have been studied to try and approach this model: a probabilistic assembly schedule [2], negative glues [15], no mismatches [5,11], and different tile shapes [7,9].

---

[1] Actually, deterministic tile assembly systems map directly to deterministic finite automata.

Due to the proximity with finite automata, a first intuition to prove the weakness of this model is that we can try to "pump" parts of an assembly between two tiles of equal type, resulting in infinite, periodic paths. However, this is not always possible, where an attempt to pump would result in a glue mismatch, which would block the growth. This conflict leads to consider three variants of this model: the case *without mismatch* where any assembly using at least twice the same tile type is pumpable, the *directed* case where conflicts can occur but where a tile assembly will always produce the same terminal assembly and the *non-directed case* where a tile assembly will not always produce the same terminal assembly.

Before this paper, a single *positive* construction was known, in which for all ε, a tileset T_ε could build multiple assemblies, all of Manhattan diameter $(2 - \varepsilon)|T_\varepsilon|$ (this means in particular that T_ε cannot build any infinite assembly). Even though that result was the first example of an *algorithmic* construction, the term "algorithm" in that case is to be taken in an extremely weak sense of a program whose running time is larger than its size. Indeed, the resulting assemblies were only a constant factor bigger than the program size, perhaps analogous to a program that calls the same function twice.

Here, we show a way to build an assembly of width $\Theta(n \log n)$ with only n different tile types, using the two dimensions to build a *"controlled loop"*:

Theorem 1. *For all $t \geq 0$, there is a tile assembly system $\mathcal{T} = (T, \sigma, 1)$, where $|\sigma| = 1$, $|T| \geq t$, and all assemblies $\alpha \in \mathcal{A}_\square[\mathcal{T}]$ are of horizontal span $w = \Theta(|T| \log |T|)$ and height less than $|T|$, and contain the same path P of width w.*

Our construction relies on non-directedness in such a crucial way that it leads us to conjecture that the expressiveness of the non-directed case is stronger than the directed one. Moreover, it also shows that there no "trivial" pumping lemma since the proof of such a result would have to deal with our construction. However, there are strong reasons to believe that 2D noncooperative tile assembly is not capable of performing Turing computation, since it is in particular not capable of simulating Turing machines inside a rectangle [13], which is the only known form of Turing computation in tile assembly.

2 Definitions and Preliminaries

These definitions are for a large part taken from [13].

2.1 Abstract Tile Assembly Model

The abstract tile assembly model was introduced by Winfree [21]. In this paper we study a restriction of this model called the temperature 1 abstract tile assembly model, or noncooperative abstract tile assembly model. For a more detailed definition of the full model, as well as intuitive explanations, see for example [16,18].

A *tile type* is a unit square with four sides, each consisting of a glue *type* and a nonnegative integer *strength*. Let T be a finite set of tile types. The sides of a tile type are respectively called north, east, south, and west.

An *assembly* is a partial function $\alpha : \mathbb{Z}^2 \dashrightarrow T$ where T is a set of tile types and the domain of α (denoted $\mathrm{dom}(\alpha)$) is connected. We let $\mathcal{A}^T$ denote the set of all assemblies over the set of tile types T. In this paper, two tile types in an assembly are said to *bind* (or *interact*, or are *stably attached*), if the glue types on their abutting sides are equal, and have strength ≥ 1. An assembly α induces a weighted *binding graph* $G_\alpha = (V, E)$, where $V = \mathrm{dom}(\alpha)$, and there is an edge $\{a, b\} \in E$ if and only if the tiles at positions a and b interact, and this edge is weighted by the glue strength of that interaction. The assembly is said to be τ-stable if every cut of G_α has weight at least τ. A *tile assembly system* is a triple $\mathcal{T} = (T, \sigma, \tau)$, where T is a finite set of tile types, σ is a τ-stable assembly called the *seed*, and $\tau \in \mathbb{N}$ is the *temperature*.

Given two τ-stable assemblies α and β, we say that α is a *subassembly* of β, and write $\alpha \sqsubseteq \beta$, if $\mathrm{dom}(\alpha) \subseteq \mathrm{dom}(\beta)$ and for all $p \in \mathrm{dom}(\alpha)$, $\alpha(p) = \beta(p)$. We also write $\alpha \to_1^{\mathcal{T}} \beta$ if we can obtain β from α by the binding of a single tile type, that is: $\alpha \sqsubseteq \beta$, $|\mathrm{dom}(\beta) \setminus \mathrm{dom}(\alpha)| = 1$ and the tile type at the position $\mathrm{dom}(\beta) \setminus \mathrm{dom}(\alpha)$ stably binds to α at that position. We say that γ is *producible* from α, and write $\alpha \to^{\mathcal{T}} \gamma$ if there is a (possibly empty) sequence $\alpha_1, \alpha_2, \ldots, \alpha_n$ where $n \in \mathbb{N} \cup \{\infty\}$, $\alpha = \alpha_1$ and $\alpha_n = \gamma$, such that $\alpha_1 \to_1^{\mathcal{T}} \alpha_2 \to_1^{\mathcal{T}} \ldots \to_1^{\mathcal{T}} \alpha_n$. A sequence of $n \in \mathbb{Z}^+ \cup \{\infty\}$ assemblies $\alpha_0, \alpha_1, \ldots$ over $\mathcal{A}^T$ is a *$\mathcal{T}$-assembly sequence* if, for all $1 \leq i < n$, $\alpha_{i-1} \to_1^{\mathcal{T}} \alpha_i$.

The set of *productions*, or *producible assemblies*, of a tile assembly system $\mathcal{T} = (T, \sigma, \tau)$ is the set of all assemblies producible from the seed assembly σ and is written $\mathcal{A}[\mathcal{T}]$. An assembly α is called *terminal* if there is no β such that $\alpha \to_1^{\mathcal{T}} \beta$. The set of all terminal assemblies of $\mathcal{T}$ is denoted $\mathcal{A}_\square[\mathcal{T}]$.

In this paper, we consider that $\tau = 1$. Thus, we make the simplifying assumption that all glue types have strength 0 or 1: it is not difficult to see that this assumption does not change the behavior of the model (if a glue type g has strength $s_g \geq 1$, in the $\tau = 1$ model then a tile with glue type g binds to a matching glue type on an assembly border irrespective of the exact value of s_g). Consider an assembly α which is producible by a tile assembly system at temperature 1, since only one glue of strength 1 is needed to stably bind a tile type to an assembly then any path of the binding graph of α can grow if it is bind to the seed. Thus at temperature 1, it is more pertinent to consider path instead of assembly. Now, we introduce definitions which are useful to study temperature 1.

2.2 Paths and Non-cooperative Self-assembly

Let T be a set of tile types. A *tile* is a pair $((x, y), t) \in \mathbb{Z}^2 \times T$ where (x, y) is a *position* and t is a tile type. Intuitively, a path is a finite or one-way-infinite simple (non-self-intersecting) sequence of tiles placed on points of $\mathbb{Z}^2$ so that each tile in the sequence interacts with the previous one, or more precisely:

Definition 2 (Path). *A path is a (finite or infinite) sequence $P = P_0 P_1 P_2 \ldots$ of tiles $P_i = ((x_i, y_i), t_i) \in \mathbb{Z}^2 \times T$, such that:*

- *for all P_j and P_{j+1} defined on P it is the case that t_j and t_{j+1} interact, and*
- *for all P_j, P_k such that $j \neq k$ it is the case that $(x_j, y_j) \neq (x_k, y_k)$.*

Whenever P is finite, i.e. $P = P_0 P_1 P_2 \ldots P_{n-1}$ for some n, n is termed the *length* of P. Note that by definition, paths are simple (or self-avoiding). We say a path P of length n is a prefix (resp. suffix) of a path P' of length n' if and only if $n \leq n'$ and for all $0 \leq i \leq n - 1, P_i = P_i'$ (resp. $P_{n-1-i} = P_{n'-1-i}'$).

Although a path is not an assembly, we know that each adjacent pair of tiles in the path sequence interact implying that the set of path positions forms a connected set in $\mathbb{Z}^2$ and hence every path uniquely represents an assembly containing exactly the tiles of the path. More formally, for a path P we define the assembly asm(P) which to a position occupied a tile of P gives the tile type of this tile (thus the domain of this function is the set of positions occupied by a tile of P). In this case, we call asm(P) a *path assembly*. A *path P is said to be producible* by some tile assembly system $\mathcal{T} = (T, \sigma, 1)$ if the assembly $(\text{asm}(P) \cup \sigma) \in \mathcal{A}[\mathcal{T}]$ is producible, and we call such a P a *producible path*. We define

$$\mathbf{P}[\mathcal{T}] = \{P \mid P \text{ is a path and } (\text{asm}(P) \cup \sigma) \in \mathcal{A}[\mathcal{T}]\}$$

to be the set of producible paths of $\mathcal{T}$.[2] If all paths of $\mathbf{P}[\mathcal{T}]$ are finite, then we can consider their last tile. If there are no free glue on the last tile of a path then it cannot grow anymore. We call such a path a *dead-end*. Note that a producible path P of $\mathbf{P}[\mathcal{T}]$ is either a dead-end or there exists a dead-end P' of $\mathbf{P}[\mathcal{T}]$ such that P is a prefix of P'. Thus If we are able to characterize all producible dead-ends then we are able to characterize $\mathbf{P}[\mathcal{T}]$.

For any path $P = P_0 P_1 P_2, \ldots$ and integer $i \geq 0$, we write pos$(P_i) \in \mathbb{Z}^2$, or $(x_{P_i}, y_{P_i}) \in \mathbb{Z}^2$, for the position of P_i and type(P_i) for the tile type of P_i. Hence if $P_i = ((x_i, y_i), t_i)$ then pos$(P_i) = (x_{P_i}, y_{P_i}) = (x_i, y_i)$ and type$(P_i) = t_i$.

Note that, since the domain of a producible assembly is a connected set in $\mathbb{Z}^2$, and since in an assembly sequence of some tile assembly system $\mathcal{T} = (T, \sigma, 1)$ each tile binding event $\beta_i \rightarrow_1^{\mathcal{T}} \beta_{i+1}$ adds a single node v to the binding graph G_{β_i} of β_i to give a new binding graph $G_{\beta_{i+1}}$, and adds at least one weight-1 edge joining v to the subgraph $G_{\beta_i} \in G_{\beta_{i+1}}$, then for any tile $((x, y), t) \in \alpha$ in a producible assembly $\alpha \in \mathcal{A}[\mathcal{T}]$, there is a edge-path (sequence of edges) in the binding graph of α from σ to $((x, y), t)$. From there, the following important fact about temperature 1 tile assembly is straightforward to see.

Observation 3. *Let $\mathcal{T} = (T, \sigma, 1)$ be a tile assembly system and let $\alpha \in \mathcal{A}[\mathcal{T}]$. For any tile $((x, y), t) \in \alpha$ there is a producible path $P \in \mathbf{P}[\mathcal{T}]$ that for some $i \in \mathbb{N}$ contains $P_i = ((x, y), t)$.*

---

[2] Intuitively, although producible paths are not assemblies, any producible path P encodes an unambiguous description of how to grow asm(P) from the seed σ, according to the order of the sequence P, to produce the assembly $\sigma \cup \text{asm}(P)$.

When referring to the relative placements of positions in the grid graph of $\mathbb{Z}^2$, we say that a position (x, y) is *east* (respectively, *west, north, south*) of another position (x', y') if $x \geq x'$ (respectively $x \leq x'$, $y \geq y'$, $y \leq y'$). A path P is east of another path P' if and only if any position occupied by a tile of P is east of all positions occupied by a tile of P'.

If two paths, or two assemblies, or a path and an assembly, share a common position we say they *intersect* at that position. Furthermore, we say that two paths, or two assemblies, or a path and an assembly, *agree* on a position if they both place the same tile type at that position and *conflict* if they place a different tile type at that position. We sometimes say that a path P is *blockable* to mean that there is another path P' (producible by the same tile assembly system as produced P) that conflicts with P.

The translation of a tile $((x, y), t)$ by a vector $\overrightarrow{v} = (x', y')$ of $\mathbb{Z}^2$ is $((x + x', y + y'), t)$ (the type of the tile is not modified while its position is translated by $\overrightarrow{v}$). The translation of a path P by $\overrightarrow{v}$, written $P + \overrightarrow{v}$, is the path Q where and for all indices i of P, $Q_i = P_i + \overrightarrow{v}$. As a convenient notation, for a path PQ composed of subpaths P and Q, when we write $PQ + \overrightarrow{v}$ we mean $(PQ) + \overrightarrow{v}$ (i.e. the translation of all of PQ by $+\overrightarrow{v}$).

The *width* of an assembly α is the number of columns on which α has at least one tile, and the *height* of α is the number of rows on which α has at least one tile.

3   The Tile Assembly System

3.1   Definition of the Tile Assembly System

Our construction relies on two parameters $k, n \in \mathbb{N}$. Also, we define the series $(h_i)_{i \geq 0}$ as:

$$h_0 = 2, h_1 = 4 \text{ and for } i \geq 2, h_i = 3h_{i-1} - h_{i-2}.$$

The tile assembly system will possess a number of tile types which is exponential according to k. The main theorem will be proven by setting wisely k and n. In this paper, we work on a zone of the $2D$ plane delimited as follow: we will only consider positions $(x, y) \in \mathbb{N}^2$ such that $0 \leq x \leq (k+1)n - 1$ and $0 \leq y \leq h_k - 1$. The seed will be made of only one tile at position $(0, 0)$ and any assembly will have a height bounded by $h_k - 1$ and a width bounded by $(k + 1)n - 1$.

The aim of this section is to define the tile assembly system $T^{(k,n)} = (\mathcal{T}^{(k,n)}, \sigma, 1)$. The path of Fig. 1 illustrates the definition of the set of tile types $\mathcal{T}^{(k,n)}$: each tile type is used exactly one time in this assembly. This path is made of five parts: one is the seed σ and the four others are represented in green, orange, blue and red. The set of tile types $\mathcal{T}^{(k,n)}$ is defined as the union of these five kinds of tile types $\mathcal{S}, \mathcal{G}, \mathcal{O}, \mathcal{B}$ and $\mathcal{R}$.

The Seed.
    Consider the tile type s with only one glue called $\mathbf{g_0}$ on its east side. We define $\mathcal{S} = \{s\}$ and the seed σ is defined as the assembly made of only the tile $((0, 0), s)$. From now on, σ will always be the seed of our tile assembly system.

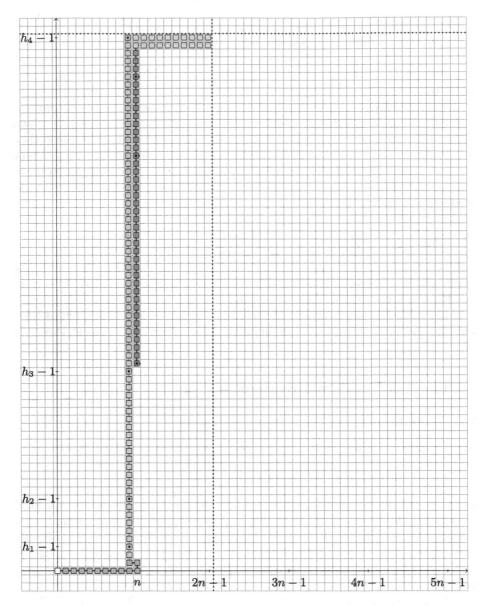

Fig. 1. In our examples, we consider $k = 4$ and $n = 10$. The seed (in white) is at position $(0,0)$ and we represent the path $S^{(4,4)} = GO^4 B^4 R^{(4,4)}$. This path is producible by $T^{(k,n)}$ and this figure contains exactly one occurrence of each tile type of $\mathcal{T}^{(k,n)}$. The height of $S^{(4,4)}$ is $h_k - 1$ and its width is $2n - 1$. The tiles of O^4 and $R^{(4,4)}$ with a glue on their east side are marked by a black dot. (Color figure online)

The green tile types.

The second kind of tile types $\mathcal{G}$ is made of $n + 2$ tile types called $g_0, g_1, \ldots, g_{n+1}$ defined as follow:

- for all $0 \leq i \leq n - 2$ the tile type g_i is made of the glue g_i on its west side and the glue g_{i+1} on its east side; moreover the tile G_i is defined as $((1+i, 0), g_i)$.
- the tile type g_{n-1} is made of the glue g_{n-1} on its west side and the glue g_n on its north side; moreover the tile G_{n-1} is defined as $((n, 0), g_{n-1})$.
- the tile type g_n is made of the glue g_n on its south side and the glue g_{n+1} on its west side; moreover the tile G_n is defined as $((n, 1), g_n)$.
- the tile type g_{n+1} is made of the glue g_{n+1} on its east side and the glue o_0 on its north side; moreover the tile G_{n+1} is defined as $((n - 1, 1), g_{n+1})$.

This set of tile types is used to hardcode the path $G = G_0 G_1 \ldots G_{n+1}$.

The orange tile types.

The third kind of tile types $\mathcal{O}$ is made of $h_k - 2$ tile types called $o_0, o_1, \ldots, o_{h_k-3}$ defined as follow. For all $0 \leq i \leq h_k - 3$ the tile type o_i is made of:

- the glue o_i on its south side;
- the glue o_{i+1} on its north side if and only if $i < h_k - 3$;
- the glue b_0 on its east side if and only if there exists $1 \leq i \leq k$ such that $i = h_i - 3$.

For all $0 \leq i \leq h_k - 3$, the tile O_i is defined as $((n - 1, 2 + i), o_i)$. This set of tile types is used to hardcode the paths $O^i = O_0 O_1 \ldots O_{h_i-3}$ for $1 \leq i \leq k$. Note that, for all $1 \leq i \leq k - 1$, the path O^i is a prefix of the path O^{i+1}.

The blue tile types.

The fourth kind of tile types $\mathcal{B}$ is made of $2n$ tile types called $b_0, b_1, \ldots, b_{2n-1}$ defined as follow:

- for all $0 \leq i \leq n - 2$ the tile type b_i is made of the glue b_i on its west side and the glue b_{i+1} on its east side; moreover the tile B_i is defined as $((n + i, 3), b_i)$;
- the tile type b_{n-1} is made of the glue b_{n-1} on its west side and the glue b_n on its south side; moreover the tile B_{n-1} is defined as $((2n - 1, 3), b_{n-1})$;
- the tile type b_n is made of the glue b_n on its north side and the glue b_{n+1} on its west side; moreover the tile B_n is defined as $((2n - 1, 2), b_n)$;
- for all $n + 1 \leq i \leq 2n - 2$, the tile type b_i is made of the glue b_i on its east side and the glue b_{i+1} on its west side; moreover the tile B_i is defined as $((3n - 1 - i, 2), b_i)$;
- the tile type b_{2n-1} is made of the glue b_{2n-1} on its east side and the glue r_0 on its south side; moreover the tile B_{2n-1} is defined as $((n, 2), b_{2n-1})$.

This set of tile types is used to hardcode the path $B^1 = B_0 B_1 \ldots B_{2n-1}$. For all $2 \leq i \leq k$, we define the path B^i as B^1 translated by $(0, h_i - 4)$.

The red tile types.

The fifth kind of tile types $\mathcal{R}$ is made of $h_k - h_{k-1} - 2$ tile types called $r_0, r_1, \ldots, r_{h_k-h_{k-1}-3}$ defined as follow. For all $0 \leq i \leq h_k - h_{k-1} - 3$, the tile type r_i is made of:

- the glue r_i on its north side;
- the glue r_{i+1} on its south side if and only if $i < h_k - h_{k-1} - 3$;
- the glue g_0 on its east side if and only if there exists $2 \leq i \leq k$ such that $i = h_i - h_{i-1} - 3$.

For all $0 \leq i \leq h_k - h_{k-1} - 3$, the tile R_i is defined as $((n, 1 - i), r_i)$. This set of tile types is used to hardcode the path R defined as $R_0 R_1 \ldots R_{h_k - h_{k-1} - 3}$. For all $1 \leq i \leq k$ and $1 \leq j \leq k$, we define the path $R^{(i,j)}$ as:

- ϵ (the empty path of length 0) if $i = 1$ or $j = 1$;
- the prefix of R of length $h_j - h_{j-1} - 2$ translated by $(0, h_i - 4)$ otherwise.

Note that, for all $1 \leq j' \leq j$, the path $R^{(i,j')}$ is a prefix of $R^{(i,j)}$.

3.2   Basic Properties

The aim of this section is to define a set of paths which characterized all the possible prefixes of a path producible by $T^{(k,n)}$. These paths are obtained by gluing together the different paths defined in Sect. 3.1. When two of these paths are glued together, we have to verify that the result of this operation is also a path. To achieve this goal, we have to check two properties. The first one is that the last tile of the first path can be glued to the first tile of the second path. The second one is that that the two paths do not intersect. We start by a first lemma which gives the positions occupied by the paths defined in Sect. 3.1. This lemma is useful to show that a path is west or north of another one and thus that these two paths do not intersect.

Fact 4. For any $k, n \in \mathbb{N}$, for any $1 \leq j \leq i \leq k$ and for any position (x, y) occupied by:

- a tile of G, we have $1 \leq x \leq n$ and $0 \leq y \leq 1$;
- a tile of O^i, we have $x = n - 1$ and $2 \leq y \leq h_i - 1$;
- a tile of B^i, we have $n \leq x \leq 2n - 1$ and $h_i - 2 \leq y \leq h_i - 1$;
- a tile of $R^{(i,j)}$, we have $x = n$ and $h_i - h_j + h_{j-1} \leq y \leq h_i - 3$ (for $j = i$, we have $h_{i-1} \leq y \leq h_i - 3$).

Now, for all $1 \leq j \leq i \leq k$, we defined $S^{(i,j)}$ as $G O^i B^i R^{(i,j)}$ and we show that these sequences of tiles are paths producible by $T^{(k,n)}$. Note that for any $1 \leq j \leq i \leq k$, $S^{(i,j)}$ is a prefix of $S^{(i,i)}$. Thus we can restrict our studies to $S^{(i,i)}$ for all $1 \leq i \leq k$.

Lemma 5. For any $k, n \in \mathbb{N}$ and for any $1 \leq i \leq k$, $S^{(i,i)}$ is a path producible by $T^{(k,n)}$.

The paths $(S^{(i,j)})_{1 \leq j \leq i \leq k}$ will be used to characterize all the paths producible by $T^{(k,n)}$. To achieve this goal, we need to know the positions of the free glues on these paths (see Fig. 1). These glues can be deduced from the free glues of $S^{(i,i)}$.

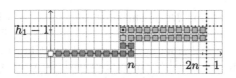

Fig. 2. The path $S^{(1,1)} = GO^1B^1 = D^1$ (for $k = 4$ and $n = 10$). The south glue of its last tile does a mismatch. Thus, this path is a dead-end. (Color figure online)

Lemma 6. For any $k, n \in \mathbb{N}$ and for any $1 \le i \le k$, the free glues of $S^{(i,i)}$ are:

- the north glue o_{h_i-2} of the tile O_{h_i-3} whose position is $(n-1, h_i-1)$ if $i < k$;
- for all $1 \le i' \le i-1$ the east glue b_0 of the tile $O_{h_{i'}-3}$ whose position is $(n-1, h_{i'}-1)$;
- for all $2 \le j \le i$ the east glue g_0 of the tile $R^{(i,i)}_{h_j-h_{j-1}-3}$ whose position is $(n, h_i - h_j + h_{j-1})$;
- the south glue $r_{h_i-h_{i-1}-2}$ of the tile $R^{(i,i)}_{h_i-h_{i-1}-3}$ whose position is (n, h_{i-1}) if $1 < i < k$.

Proof. Pretty straightforward, except for the special case where $i = 1$ (see Fig. 2) the last tile of $GO^1B^1R^{(1,1)}$ is the last tile of B^1. Nevertheless, the south glue of the tile B^1_{2n-1} whose position is $(n, 2)$ is not free because of a mismatch with the tile G_n whose position is $(n, 1)$ and which has no glue on its north side.

A corollary of this lemma is that $S^{(1,1)}$ is a dead-end (see Fig. 2). Also, for all $2 \le j \le i \le k$, let $\overrightarrow{v}^{(i,j)}$ be the vector $(n, h_i - h_j + h_{j-1})$ and since the last tile of $S^{(i,j)}$ is $(n, h_i - h_j + h_{j-1})$ and since the type of its east east glue is g_0 then for all P producible by $T^{(k,n)}$, if $S^{(i,j)}(G_0 + \overrightarrow{v}^{(i,j)})$ is a prefix of P then P can be written $S^{(i,j)}P'$ where $P' - \overrightarrow{v}^{(i,j)}$ is a path producible by $T^{(k,n)}$.

3.3 Analysis of the Prefixes

Now, remark that the only tile with a glue g_0 on its west side is G_0 then any path producible by $T^{(k,n)}$ begins by a tile G_0. In fact, this reasoning can be done for any glue and direction. Thus for any path P and any $0 \le i < |P| - 1$, if we know the position and the tile type of P_i and the position of P_{i+1} then we can deduce the tile type of P_{i+1}. Thus, consider two paths P and P' which are producible by $T^{(k,n)}$, then they share a common prefix until they split away. They can split away only at a tile with at least three glues on its sides. From Lemma 6, we have the following fact:

Fact 7. For any $k, n \in \mathbb{N}$ and for any path P producible by $T^{(k,n)}$, either P is a prefix of GO^k or there exists $1 \le i \le k$ such that $GO^iB^i_0$ is a prefix of P.

There are only k different prefixes for a producible path with is large enough. Now, let's look at how these different prefixes can grow. For all $1 \le i \le k$, by attaching tiles to the end of $GO^iB^i_0$, this path will always grow into GO^iB^i.

For all $1 \leq i \leq k$, we define D^i as the path obtained by attaching red tiles to GO^iB^i until it is no more possible to do so (see Fig. 3). For the special case $i = k$, we have $D^k = S^{(k,k)}$ since we run out of red tiles. For the other special case $i = 1$, the path D^1 is $S^{(1,1)}$ since this path is a dead-end then it is not possible to add any red tiles (see Fig. 2). For the general case $1 < i < k$, D^i is a dead-end since its last tile is $((n,1), r_{h_i-4})$ and its only free glue is the south one which does a mismatch with the tile $G_n = ((n,1), g_n)$. In all cases, there are only $i - 1$ red tiles with a free glue on their east side (see Lemma 6). Thus if a path producible by $T^{(k,n)}$ is not a prefix of D^i then it has to use on these free glues. The previous remarks are summarized in the following fact.

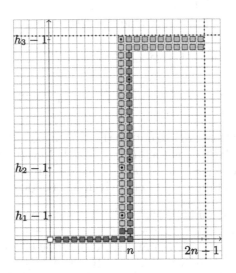

Fig. 3. The path D^3 for $k = 4$ and $n = 10$. This path is obtained by attaching red tiles to GO^3B^3 until a conflict occurs with G. For all $1 \leq i < k$, the path D^i is a dead-end. The path D^k is a special case since it is equal to $S^{(k,k)}$ (see Fig. 1) whose last tile has a free glue on its east side. (Color figure online)

Fact 8. For any $k, n \in \mathbb{N}$ and for any path P producible by $T^{(k,n)}$, there exists $1 \leq i \leq k$ such that either:

- P is a prefix of D^i;
- there exists $2 \leq j \leq i$ such that $S^{(i,j)}(G_0 + \overrightarrow{v}^{(i,j)})$ is a prefix of P.

Moreover, for all $1 \leq i < k$, D^i is a dead-end.

Now, we have obtained $k-1$ different prefixes which are dead-end and $(k-1)^2$ prefixes which are not. Now, let's look at how these different prefixes can grow. Consider $2 \leq j \leq i \leq k$ and by attaching tiles to the end of $S^{(i,j)}$, this path will always grow into $S^{(i,j)}(G + \overrightarrow{v}^{(i,j)})$ (this assembly is producible by $T^{(k,n)}$ since $G + \overrightarrow{v}^{(i,j)}$ is east of O^i and $R^{(i,j)}$, north of G and south of B^i (see Lemma 4)).

The last tile of this path is $((2n-1, h_i - h_j + h_{j-1} + 1), g_{n+1})$ and this path can keep growing to the north by attaching orange tiles at its end. We define the path $D'^{(i,j)}$ as the path obtained by attaching orange tiles to $S^{(i,j)}(G + \overrightarrow{v}^{(i,j)})$ until it is no more possible to do so (see Fig. 4). The last tile of this path is $((2n-1, h_i - 3), o_{h_j - h_{j-1} - 5})$ since it not possible to add an orange tile because of a mismatch with the tile $B_n^i = ((2n-1, h_i - 2), b_n)$. Moreover if $j > 2$, this path is a dead-end because the last tile of this path has no east glue. Also, $GO^j + \overrightarrow{v}^{(i,j)}$ cannot be a prefix of P' and by applying Lemma 7 to $P' - \overrightarrow{v}^{(i,j)}$ we obtain that either P is a prefix of $D'^{(i,j)}$ or there exists $1 \le i' < j$ such that $S^{(i,j)}(GO^{i'} B_0^{i'}) + \overrightarrow{v}^{(i,j)}$ is a prefix of P. These observations are summarized in the following fact.

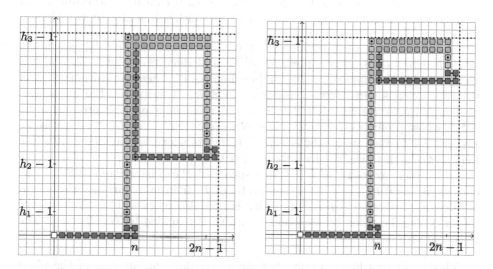

Fig. 4. The path $D'^{(3,3)}$ (on the left) and $D'^{(3,2)}$ (on the right) for $k = 4$ and $n = 10$. These paths are obtained by attaching blue and orange tiles to $S^{(3,3)}$ and $S^{(3,2)}$ until a conflict occurs with B^3. For all $2 < j \le i \le k$, the path $D'^{(i,j)}$ is a dead-end. (Color figure online)

Fact 9. For any $k, n \in \mathbb{N}$ and for any path P producible by $T^{(k,n)}$, there exists $1 \le i \le k$ such that either:

- P is a prefix of D^i;
- there exists $2 \le j \le i$ such that either:
  - P is prefix of $D'^{(i,j)}$;
  - there exists $1 \le i' < j$ such that $S^{(i,j)}(GO^{i'} B_0^{i'} + \overrightarrow{v}^{(i,j)})$ is a prefix of P.

Moreover, for all $2 < j \le i \le k$, $D'^{(i,j)}$ is a dead-end.

3.4   Analysis of the Tile Assembly System

Consider a path P producible by $T^{(k,n)}$ then Fact 9 gives us a hint to the structure of P. Indeed, this path is made of several paths $S^{(i_1,j_1)}, S^{(i_2,j_2)}, S^{(i_3,j_3)}, \ldots$ which are glued together up to some translations (see Fig. 5). The index i (resp. j) represents which east glue of an orange (resp. red) tile is used. Moreover, Fact 9 also implies that $i_1 < i_2 < i_3 < \ldots$ and thus eventually one the three kinds of dead-end $S^{(1,1)}, D^i, D'^{(i,j)}$ (for some $2 \leq j \leq i \leq k$) will grow up to some translation and the path will become a dead-end too. We use this sketch of proof to show that the width and height of any producible path is bounded.

Lemma 10. For any $k, n \in \mathbb{N}$ and for any path P producible by $T^{(k,n)}$, if there exists $1 \leq i \leq k$ such that $GO^i B_0^i$ is a prefix of P then P_0 is its southernmost and westernmost tile and its height is $h_i - 1$ and its width is bounded by $(i+1)n - 1$.

Proof. This proof is done by recurrence on i. For $i = 1$, if $GO^1 B_0^1$ is a prefix of P then by Fact 8, P is a prefix of $S^{(1,1)}$ and thus its height is $h_1 - 1 = 3$ and its width is bounded by $2n - 1$ (see Lemma 4). Then the initialization of the recurrence is true. Now, consider $1 \leq i \leq k - 1$ such that the hypothesis of recurrence is true for all $1 \leq i' \leq i$ and consider a path P producible by $T^{(k,n)}$ whose prefix is $GO^{i+1} B_0^{i+1}$. If P is a prefix of D^{i+1} or $D'^{(i+1,j)}$ (for some $1 \leq j \leq i + 1$) then its height is $h_{i+1} - 1$ and its width is bounded by $2n$ (see Lemma 4). In this case, the hypothesis of recurrence is true for $i + 1$. Otherwise by Fact 9, there exists $1 \leq i' < j \leq i+1$ such that P can be written as $S^{(i+1,j)} P'$ and where $(GO^{i'} B_0^{i'}) + \overrightarrow{v}^{(i+1,j)}$ is a prefix of P' then by recurrence the height of $P' - \overrightarrow{v}^{(i+1,j)}$ is $h_{i'} - 1$ and its width is bounded by $(i' + 1)n - 1$. Thus the height of P is bounded by h_{i+1} and its width is bounded by $(i + 2)n - 1$. The recurrence is true for $i + 1$.

This result and Fact 7 imply that we are working on an area of the plane delimited by $0 \leq x \leq (k+1)n - 1$ and $0 \leq y \leq h_{n+1}$ and that for any path P producible by $T^{(k,n)}$, its first tile is the westernmost and southernmost one.

Corollary 11. For any $k, n \in \mathbb{N}$ and for any terminal assembly $\alpha \in \mathcal{A}_\square[T^{(k,n)}]$, the height of α is bounded by h_{n+1} and its width is bounded by $(k+1)n - 1$.

Now, we aim to assemble the largest path possible (see Fig. 6). Note that, for all $1 \leq i \leq k$, the last tile of the path $S^{(i,i)}$ is at distance n to the east of its first tile (this property is due to the definition of the green tiles). Thus, if we manage to glue k paths together, we will obtain a path whose width is $(k+1)n - 1$. This is the path we are looking for. Formally, we define the path L^1 as $S^{(1,1)}$ and for all $2 \leq i \leq k$, we define L^i as $S^{(i,i)}(L^{i-1} + \overrightarrow{v}^{(i,i)})$. We remind that $\overrightarrow{v}^{(i,i)} = (n, h_{i-1})$. Now, we show that this sequence of tiles is a path producible by $T^{(k,n)}$.

Lemma 12. For all $1 \leq i \leq k$, L^i is a path producible by $T^{(k,n)}$ of width $(i+1)n - 1$.

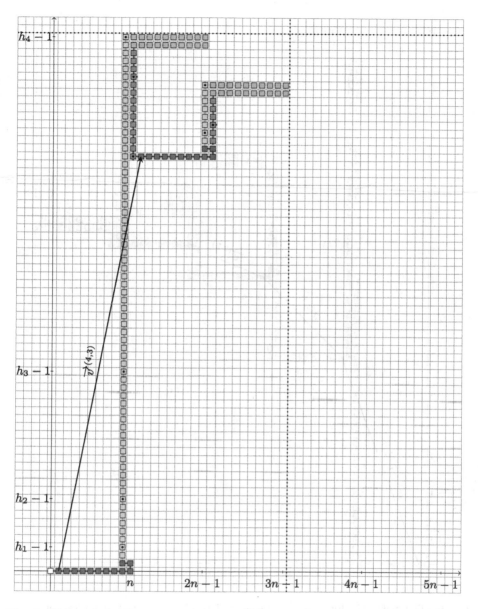

Fig. 5. This path is the concatenation of $S^{(4,3)}$ and D^2 up to some translation. It is producible by $T^{(k,n)}$ (for $k = 4$ and $n = 10$) and it is a dead-end since its suffix is D^2 up to some translation. (Color figure online)

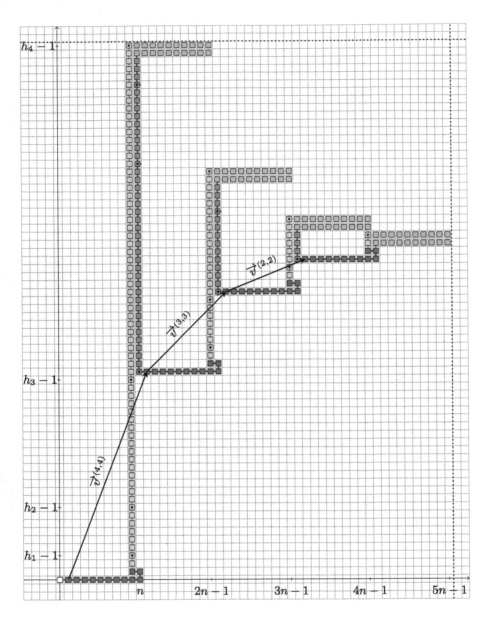

Fig. 6. The path L^4, for $k = 4$ and $n = 10$. This path is the largest one producible by $T^{(k,n)}$ and it appears in any terminal assembly. It is made of the paths $S^{(4,4)}, S^{(3,3)}, S^{(2,2)}$ and $S^{(1,1)}$ which are glued together up to some translations. (Color figure online)

Proof. This proof is done by recurrence on i. By Lemma 5, $S^{(1,1)}$ is producible by $T^{(k,n)}$ and the position of the tile B_{n-1}^1 is $(2n-1,3)$ Thus the initialization of the recurrence is true. Now, consider $1 \leq i < k$ such that the recurrence is true for i. By definition, $GO^i B_0^i$ is a prefix of L^i. Then by Lemma 10, for any position $(x,y) \in \mathbb{N}^2$ occupied by a tile of $L^i + \overrightarrow{v}^{(i+1,i+1)}$, we have $n+1 \leq x \leq (i+2)n-1$ and $h_i \leq y \leq 2h_i-1$. By Lemma 4, $L^i + \overrightarrow{v}^{(i+1,i+1)}$ is west of O^{i+1} and $R^{(i+1,i+1)}$, north of G and south of B^{i+1} and thus $L^i + \overrightarrow{v}^{(i,i)}$ neither intersects $S^{(i+1,i+1)}$ nor the seed. By recurrence L^i is a path producible by $T^{(k,n)}$ and thus $S^{(i+1,i+1)}(L^i + \overrightarrow{v}^{(i,i)})$ is also a path producible by $T^{(k,n)}$. Finally, the width of $S^{(i,i)}$ is $(i+1)n-1$ by recurrence and then the width of L^{i+1} is $(i+2)n-1$.

Lemma 13. *For any $k,n \in \mathbb{N}$ and for any terminal assembly $\alpha \in \mathcal{A}_\square[T^{(k,n)}]$, L^k is a subassembly of α.*

Proof. Consider the following hypothesis of recurrence for $1 \leq i \leq k$: "Consider a path P producible by $T^{(k,n)}$, if there exists $1 \leq i' \leq i$ such that $GO^{i'} B_0^{i'}$ is prefix a P then there is no conflict between L^i and P". By Fact 8, if $GO^1 B_0^1$ is a prefix of P then P is a prefix of $S^{(1,1)} = L^1$ and the recurrence is true for $i=1$. Now suppose that the recurrence is true for $1 \leq i < k$. By definition, $GO^i B_0^i$ is a prefix of L^i. Then by Lemma 10, for any position $(x,y) \in \mathbb{N}^2$ occupied by a tile of $L^i + \overrightarrow{v}^{(i+1,i+1)}$, we have $h_i \leq y \leq 2h_i - 1$. Consider a path P producible by $T^{(k,n)}$ such that $GO^{i'} B_0^{i'}$ is prefix of P for some $1 \leq i' \leq i$. To prove the recurrence for $i+1$, we have to study five cases which are illustrated in Fig. 7.

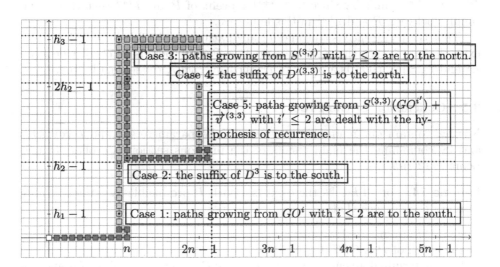

Fig. 7. Graphical representation of the step of recurrence of the proof of Lemma 13 for $k = 4, n = 10$ and $i = 3$. We represent here a prefix of L^3, note that the remaining suffix of L^3 is in the stripe defined by $h_2 \leq y \leq 2h_2 - 1$. If a path producible by $T^{(k,n)}$ creates a conflict with L^3 then it has to fork from this prefix by one of its free glue. We represent the five possible cases and give the main argument to deal with each case. (Color figure online)

Case 1: if $i' < i$ then by Lemma 10, P is south of $L^i + \overrightarrow{v}^{(i+1,i+1)}$. Also, P is south of the suffix of $S^{(i+1,i+1)}$ obtained by removing the prefix $GO^{i'}$ to $S^{(i+1,i+1)}$. Thus, there is no conflict between P and L^{i+1} in this case.

Case 2: suppose that P is a prefix of D^{i+1} then $S^{(i+1,i+1)}$ is a prefix of both L^{i+1} and D^{i+1}. Also, $L^i + \overrightarrow{v}^{(i+1,i+1)}$ is north of the suffix of D^{i+1} obtained by removing the prefix $S^{(i+1,i+1)}$ to D^{i+1}. Thus, there is no conflict between P and L^{i+1} in this case.

Case 3: if there exists $1 \le j < i+1$ such that $S^{(i,j)}(G_0 + \overrightarrow{v}^{(i,j)})$ is a prefix of P. Then, $S^{(i+1,j)}$ is a prefix of both L^{i+1} and P. Moreover, P can be written as $S^{(i+1,j)}P'$ where $P' - \overrightarrow{v}^{(i,j)}$ is producible by $T^{(k,n)}$. Thus by Lemma 10, P'_0 is the southernmost tile of P'. Then for any position $(x,y) \in \mathbb{N}^2$ occupied by a tile of P', we have $y \ge h_{i+1} - h_j + h_{j-1} \ge h_{i+1} - h_i + h_{i-1} \ge 2h_i$. Then P' is north of $L^i + \overrightarrow{v}^{(i+1,i+1)}$. Also, P' is north of the suffix of $S^{(i+1,i+1)}$ obtained by removing the prefix $S^{(i+1,j)}$ to $S^{(i+1,i+1)}$. Thus, there is no conflict between P and L^{i+1} in this case.

Case 4: suppose that P is prefix of $D'^{(i+1,i+1)}$ then $S^{(i+1,i+1)}(GO^i) + \overrightarrow{v}^{(i+1,i+1)}$ is a prefix of both $D'^{(i+1,i+1)}$ and L^{i+1}. Also, $L^i + \overrightarrow{v}^{(i+1,i+1)}$ is south of the suffix of $D'^{(i+1,i+1)}$ obtained by removing the prefix $S^{(i+1,i+1)}(GO^i) + \overrightarrow{v}^{(i+1,i+1)}$ to $D'^{(i+1,i+1)}$. Thus, there is no conflict between P and L^{i+1} in this case.

Case 5: if P does no match the previous cases, then by Fact 9, there exists $1 \le i' < i+1$ such that P can be written as $S^{(i+1,i+1)}P'$ where $(GO^{i'}) + \overrightarrow{v}^{(i+1,i+1)}$ is a prefix of P'. Since $i < i+1$ then by recurrence, there is no conflict between $P' - \overrightarrow{v}^{(i+1,i+1)}$ and L^i. Since $S^{(i+1,i+1)}$ is prefix of P and L^{i+1} then there is no conflict between P and L^{i+1} in this final case.

Thus the recurrence is true and for any path P producible by $T^{(k,n)}$, if there exists $1 \le i \le k$ such that $GO^i B_0^i$ is prefix a P then there is no conflict between L^k and P. Since GO^k is a prefix of L^k then by Fact 7, the lemma is true.

3.5   Conclusion of the Proof

Now, we obtain our main result by setting the parameters k and n correctly and by combining the main results of the previous section (Corollary 11 and Lemmas 12 and 13).

Lemma 14. For all $k \in \mathbb{N}$, there exists a path P producible by a tile assembly system $(\mathcal{T}, \sigma, 1)$ such that:

– $|\mathcal{T}| = 8h_k - h_{k-1} - 1$;
– the width of P is $\Theta(|\mathcal{T}| \log_3(|\mathcal{T}|))$;
– for any terminal assembly $\alpha \in \mathcal{A}_\square[\mathcal{T}]$, $\text{asm}(P)$ is a subassembly of α and the width and height of α are bounded by $\Theta(|\mathcal{T}| \log_3(|\mathcal{T}|))$.

Proof. Consider $k, n \in \mathbb{N}$ and the tile assembly system $T^{(k,n)}$, its set of tile types is made of 1 tiles types for the seed, $n+2$ green tiles types, $h_k - 2$ orange tile types, $2n$ blue tile types, $h_k - h_{k-1} - 2$ red tile types. Thus, we have $|\mathcal{T}^{(k,n)}| = 3n + 2h_k - h_{k-1} - 1$. By setting $n = 2h_k$, we obtain that $|\mathcal{T}^{(k,n)}| = 8h_k - h_{k-1} - 1 \le 4n$.

Moreover, a simple induction shows that for any $k \geq 3$, we have $h_k \leq 3^k$. Thus, if $k \geq 3$ then $k \geq \log_3(n) - 1$. Now, this theorem is a corollary of Corollary 11, Lemmas 12 and 13.

4 Open Questions

Three questions arise from this article. The first one is to improve our construction by increasing its width or to prove its optimality. The second question is to characterize the tile assembly systems whose terminal assemblies are finite: is it possible to build larger finite terminal assembly without relying on this method? The third question is to investigate if this construction can be achieved in the directed case: either the directed case is more complex than initially though or there is a difference in complexity between the directed and non-directed case.

Acknowledgments. We would like to thank Damien Woods for invaluable discussions, comments, and suggestions.

A Make Your Own Large Paths

A python script is available to generate a LATEX document with the same kind of terminal assembly as in Fig. 6. For any value of the two integers $k \geq 1$ and $n \geq 1$ used in this proof, this script is meant to be called in the following way:

```
$ python positive.py k n
```
For example, the drawing in Fig. 6 was produced with:
```
$ python positive.py 3 10
```
The script is available on https://github.com/P-E-Meunier/largepaths.

References

1. Cannon, S., et al.: Two hands are better than one (up to constant factors). In: STACS: Proceedings of the Thirtieth International Symposium on Theoretical Aspects of Computer Science, pp. 172–184. LIPIcs (2013). arxiv preprint: arXiv:1201.1650
2. Cook, M., Fu, Y., Schweller, R.T.: Temperature 1 self-assembly: deterministic assembly in 3D and probabilistic assembly in 2D. In: SODA: Proceedings of the 22nd Annual ACM-SIAM Symposium on Discrete Algorithms, pp. 570–589 (2011). arxiv preprint: arXiv:0912.0027
3. Demaine, E.D., et al.: One tile to rule them all: simulating any tile assembly system with a single universal tile. In: Esparza, J., Fraigniaud, P., Husfeldt, T., Koutsoupias, E. (eds.) ICALP 2014. LNCS, vol. 8572, pp. 368–379. Springer, Heidelberg (2014). https://doi.org/10.1007/978-3-662-43948-7_31. arxiv preprint: arXiv:1212.4756
4. Doty, D., Lutz, J.H., Patitz, M.J., Schweller, R.T., Summers, S.M., Woods, D.: The tile assembly model is intrinsically universal. In: FOCS: Proceedings of the 53rd Annual IEEE Symposium on Foundations of Computer Science, pp. 439–446. IEEE, October 2012. arxiv preprint: arXiv:1111.3097

5. Doty, D., Patitz, M.J., Summers, S.M.: Limitations of self-assembly at temperature 1. Theor. Comput. Sci. **412**(1–2), 145–158 (2011). arxiv preprint: arXiv:0906.3251
6. Fekete, S.P., Hendricks, J., Patitz, M.J., Rogers, T.A., Schweller, R.T.: Universal computation with arbitrary polyomino tiles in non-cooperative self-assembly. In: SODA: ACM-SIAM Symposium on Discrete Algorithms, pp. 148–167. SIAM (2015). http://arxiv.org/abs/1408.3351
7. Fekete, S.P., Hendricks, J., Patitz, M.J., Rogers, T.A., Schweller, R.T.: Universal computation with arbitrary polyomino tiles in non-cooperative self-assembly. In: Indyk, P. (ed.) Proceedings of the Twenty-Sixth Annual ACM-SIAM Symposium on Discrete Algorithms, SODA 2015, San Diego, CA, USA, 4–6 January 2015, pp. 148–167. SIAM (2015). https://doi.org/10.1137/1.9781611973730.12
8. Gilbert, O., Hendricks, J., Patitz, M.J., Rogers, T.A.: Computing in continuous space with self-assembling polygonal tiles. In: SODA: ACM-SIAM Symposium on Discrete Algorithms, pp. 937–956. SIAM (2016). arxiv preprint: arXiv:1503.00327
9. Gilbert, O., Hendricks, J., Patitz, M.J., Rogers, T.A.: Computing in continuous space with self-assembling polygonal tiles (extended abstract). In: Krauthgamer, R. (ed.) Proceedings of the Twenty-Seventh Annual ACM-SIAM Symposium on Discrete Algorithms, SODA 2016, Arlington, VA, USA, 10–12 January 2016, pp. 937–956. SIAM (2016). https://doi.org/10.1137/1.9781611974331.ch67
10. Hendricks, J., Patitz, M.J., Rogers, T.A., Summers, S.M.: The power of duples (in self-assembly): it's not so hip to be square. In: Cai, Z., Zelikovsky, A., Bourgeois, A. (eds.) COCOON 2014. LNCS, vol. 8591, pp. 215–226. Springer, Cham (2014). https://doi.org/10.1007/978-3-319-08783-2_19. arxiv preprint: arXiv:1402.4515
11. Maňuch, J., Stacho, L., Stoll, C.: Two lower bounds for self-assemblies at temperature 1. J. Comput. Biol. **17**(6), 841–852 (2010)
12. Meunier, P.É., Patitz, M.J., Summers, S.M., Theyssier, G., Winslow, A., Woods, D.: Intrinsic universality in tile self-assembly requires cooperation. In: SODA: Proceedings of the ACM-SIAM Symposium on Discrete Algorithms, pp. 752–771 (2014). arxiv preprint: arXiv:1304.1679
13. Meunier, P., Woods, D.: The non-cooperative tile assembly model is not intrinsically universal or capable of bounded Turing machine simulation. In: STOC: Proceedings of the 49th Annual ACM SIGACT Symposium on Theory of Computing, pp. 328–341 (2017)
14. Padilla, J.E., Patitz, M.J., Schweller, R.T., Seeman, N.C., Summers, S.M., Zhong, X.: Asynchronous signal passing for tile self-assembly: fuel efficient computation and efficient assembly of shapes. Int. J. Found. Comput. Sci. **25**(4), 459–488 (2014). arxiv preprint: arxiv:1202.5012
15. Patitz, M.J., Schweller, R.T., Summers, S.M.: Exact shapes and turing universality at temperature 1 with a single negative glue. In: Cardelli, L., Shih, W. (eds.) DNA 2011. LNCS, vol. 6937, pp. 175–189. Springer, Heidelberg (2011). https://doi.org/10.1007/978-3-642-23638-9_15. arxiv preprint: arXiv:1105.1215, http://dl.acm.org/citation.cfm?id=2042033.2042050
16. Rothemund, P.W.K.: Theory and experiments in algorithmic self-assembly. Ph.D. thesis, University of Southern California, December 2001
17. Rothemund, P.W.K.: Folding DNA to create nanoscale shapes and patterns. Nature **440**(7082), 297–302 (2006). https://doi.org/10.1038/nature04586
18. Rothemund, P.W.K., Winfree, E.: The program-size complexity of self-assembled squares (extended abstract). In: STOC: Proceedings of the Thirty-Second Annual ACM Symposium on Theory of Computing, pp. 459–468. ACM, Portland (2000). http://doi.acm.org/10.1145/335305.335358

19. Soloveichik, D., Winfree, E.: Complexity of self-assembled shapes. SIAM J. Comput. **36**(6), 1544–1569 (2007)
20. Thubagere, A.J., et al.: A cargo-sorting DNA robot. Science **357**(6356), eaan6558 (2017)
21. Winfree, E.: Algorithmic self-assembly of DNA. Ph.D. thesis, California Institute of Technology, June 1998
22. Winfree, E.: Simulations of computing by self-assembly. Technical report, Caltech CS TR:1998.22, California Institute of Technology (1998)
23. Yurke, B., Turberfield, A.J., Mills, A.P., Simmel, F.C., Neumann, J.L.: A DNA-fuelled molecular machine made of DNA. Nature **406**(6796), 605–608 (2000)

Simulation of Programmable Matter Systems Using Active Tile-Based Self-Assembly

John Calvin Alumbaugh[1], Joshua J. Daymude[2] (iD), Erik D. Demaine[3],
Matthew J. Patitz[1(✉)] (iD), and Andréa W. Richa[2]

[1] Department of Computer Science and Computer Engineering,
University of Arkansas, Fayetteville, AR, USA
`mpatitz@self-assembly.net`
[2] Computer Science, CIDSE, Arizona State University, Tempe, AZ, USA
`{jdaymude,aricha}@asu.edu`
[3] MIT Computer Science and Artificial Intelligence Laboratory,
Cambridge, MA, USA
`edemaine@mit.edu`

Abstract. Self-assembly refers to the process by which small, simple components mix and combine to form complex structures using only local interactions. Designed as a hybrid between tile assembly models and cellular automata, the *Tile Automata (TA) model* was recently introduced as a platform to help study connections between various models of self-assembly. However, in this paper we present a result in which we use TA to simulate arbitrary systems within the *amoebot model*, a theoretical model of programmable matter in which the individual components are relatively simple state machines that are able to sense the states of their neighbors and to move via series of expansions and contractions.

We show that for every amoebot system, there is a TA system capable of simulating the local information transmission built into amoebot particles, and that the TA "macrotiles" used to simulate its particles are capable of simulating movement (via attachment and detachment operations) while maintaining the necessary properties of amoebot particle systems. The TA systems are able to utilize only the local interactions of state changes and binding and unbinding along tile edges, but are able to fully simulate the dynamics of these programmable matter systems.

Keywords: Programmable matter · Simulation · Self-assembly · Tile Automata · Amoebot model

1 Introduction

Theoretical models of self-assembling systems are mathematical models that allow for the exploration of the limits of bottom-up construction and self-

Daymude and Richa are funded in part by the National Science Foundation under awards CCF-1422603, CCF-1637393, and CCF-1733680. Alumbaugh and Patitz are funded in part by National Science Foundation Grants CCF-1422152 and CAREER-1553166.

C. Thachuk and Y. Liu (Eds.): DNA 25, LNCS 11648, pp. 140–158, 2019.
https://doi.org/10.1007/978-3-030-26807-7_8

assembly via simple (usually square) tiles. There are a wide variety of tile-based models of self-assembly (e.g., [13,14,16,17,20–22]), each with differing constraints and dynamics, resulting in great variations in the relative powers between systems. One of the easiest ways to evaluate their relationships is to use notions of simulation to attempt to simulate one model by another, and this has led to the creation of a "complexity hierarchy" of self-assembly models and categories of systems [8,12,15,18,23].

Another category of theoretical models attempts to capture the dynamics of so-called *programmable matter*, in which small and simple, but dynamic and mobile, components are able to interact with each other to form structures, perform tasks and computations, etc. [7,24].

This paper attempts to bridge the divide between these categories of models, showing how self-assembling tiles can mimic the behaviors of programmable matter. Specifically, we demonstrate how the recently introduced *Tile Automata (TA) model* [5] can be used to simulate the *amoebot model* [7]. In the TA model, the fundamental components are unit square tiles which form structures by attaching and forming bonds, and can also change states based on their own states and those of their neighbors, causing them to be able to form new bonds or to remove existing bonds. The basic components in the amoebot model are *particles* which can also change their states based on the current states of themselves and their neighbors, but which can also move via series of expansions and contractions. While the components of both models rely only upon local information and communication, the design goals of their systems tend to differ fundamentally. The main goal of TA systems is to self-assemble into target structures, but amoebot systems have been used to solve system-level problems of movement and coordination (e.g., shape formation [9], object coating [10], leader election [6], gathering [4], bridging gaps [2], etc.). We present a construction in which constant-sized assemblies of TA tiles, called *macrotiles*, assemble and disassemble following the rules of the TA model and are able to simulate the behaviors of individual amoebot particles. Via carefully designed processes of building and breaking apart assemblies, they are collectively able to correctly simulate the full dynamics of amoebot systems. We thus show how the dynamics of systems of self-assembling tiles with the ability to form and break bonds can be harnessed to faithfully simulate the dynamics of collections of programmable matter particles capable of local communication and motion. Not only does this provide a way to connect and leverage existing results across models, this also provides a new paradigm for designing systems to accomplish the goals of programmable matter. It additionally allows amoebots to serve as a higher-level abstraction for designing systems exhibiting complex behaviors of programmable matter but with a translation to implementation in TA.

The paper is organized as follows. Section 2 presents a high-level definition of the TA model, and Sect. 3 provides a full mathematical definition for the amoebot model. (We note that this is the first full mathematical definition for the amoebot model and thus is also a contribution of this paper.) Section 4 gives the formal definition, preliminaries, and overview of the simulation of the amoebot model by TA, while Sect. 5 gives more of its details. A brief discussion and conclusion are given in Sect. 6. Due to space constraints, not all details are contained in this version. Please see [1] for an additional Technical Appendix containing a more rigorous definition of the TA model, as well as low-level technical details about the construction.

2   The Tile Automata Model

The Tile Automata model seeks to connect and evaluate the differences between some of the seemingly disparate models of tile-based self assembly by combining components of the *Two Handed Assembly Model* (2HAM) of self-assembly with a rule set of local state changes that are similar to asynchronous cellular automata. This section provides an overview of the TA model which is sufficient for the purposes of this paper, however a more thorough and detailed definition of the TA model is available in the technical appendix, which is based on [5] (Fig. 1).

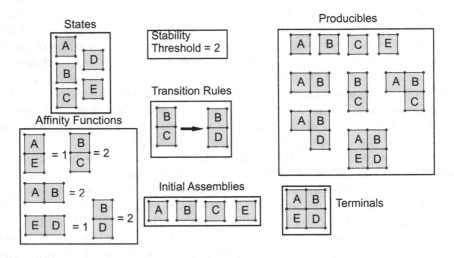

Fig. 1. Example of a TA system. The five components that define a TA system constitute the left and middle columns of this figure, while the rightmost boxes indicate producible and terminal assemblies.

The Tile Automata Model has many similarities with other tile based self assembly systems. Tiles, the fundamental units of this model that interact with one another, use only local information, in this case the *state* of their neighbors. Tiles exist as a stateful unit square centered on a point on the square lattice over the integers in two dimensions, so that a tile's coordinates $(x, y) \in \mathbb{Z}^2$. Tiles may form bonds with adjacent neighbors via attaching to one another according to the *affinity function*, which defines a set of two states and either a vertical or horizontal relative orientation (denoted as $\perp$ and $\vdash$, respectively) as well as an attachment strength. A connection between tiles or groups of connected tiles must have the property of τ stability to persist. Every TA system has defined an integer *stability threshold* or τ that represents the minimum strength bond with which tiles must be bound in order to be τ stable. Two adjacent tiles of states s, s', with the tile of state s directly to the right of the tile of state s', will form an attachment if there exists a rule in the affinity function $(s' \vdash s \geq \tau)$. An *assembly* is a τ stable connected set of TA tiles, with the property that there exists no way to separate the tiles without breaking bonds of at least τ strength. Further, a pair of tiles may *transition* according to a transition rule that takes as input two adjacent tiles (oriented by either $\perp$ or $\vdash$) and outputs new states for those tiles. So the tiles in our example of $s' \vdash s$ may transition to states $t \vdash s$ if there exists a rule in the set of transition rules provided in the definition of a TA system of the form $(s', s, t, s, \vdash)$, where (s', s) are the input states, (t, s) are the output states, and $\vdash$ is their relative orientation.

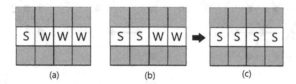

Fig. 2. Depiction of signal S being passed down a wire. The W tiles represent wires in their default state, and the grey tiles above and below the wire are filler tiles. Starting in state in (a) and a transition rule (SW $\vdash$ SS), the signal propagates down the wire in (b) and (c).

2.1   Wire Transmission

One of the most useful aspects of the Tile Automata model is the tiles' ability to transition states based on local information. This capability makes communication from one group of tiles to another, non-adjacent group easy, with a structure we will call a *wire*. A wire in TA is a contiguous line of tiles from one group of tiles to another, usually surrounded by inert filler tiles so as to avoid interference with the signal being transmitted. (See Fig. 2 for an example.)

3   The Amoebot Model

Introduced in [11], the amoebot model is an abstract computational model of *programmable matter*, a substance that can change its physical properties based on user input or stimuli from its environment. The amoebot model envisions programmable matter as a collection of individual, homogeneous computational elements called *particles*. In what follows, we extend the exposition of the model in [7] to the level of formality needed for our simulation.

Any structure a particle system can form is represented as a subgraph of an infinite, undirected graph $G = (V, E)$ where V is the set of positions a particle can occupy and E is the set of all atomic movements a particle can make. Each node in V can be occupied by at most one particle at a time. This work further assumes the *geometric amoebot model* where $G = G_\Delta$, the triangular lattice with nearest neighbor connectivity (see Fig. 3a). This lattice is preferred for work in the 2D plane, as it allows for a maximum of nearest neighbor connectivity for particles moving step wise around the perimeter of the particle swarm. Particles attempting to move around a "corner" of a particle swarm risk disconnection with the neighborhood implied by nearest neighbor connectivity on the square lattice. Each particle occupies either a single node in V (i.e., it is *contracted*) or a pair of adjacent nodes in V (i.e., it is *expanded*), as in Fig. 3b. Two particles occupying adjacent nodes of G_Δ are *neighbors*. We further will define a group of particles as a *particle system*.

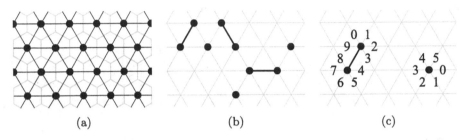

(a)   (b)   (c)

Fig. 3. (a) A section of the triangular lattice G_Δ (black) and its dual, the hexagonal tiling (gray). (b) Expanded and contracted particles (black dots) on G_Δ (gray lattice). Particles with a black line between their nodes are expanded. (c) Two particles with different orientations. The expanded particle's tail port would be 6 if its head were the upper node; the contracted particle's tail port is ε.

Each particle keeps a collection of *ports*—one for each edge incident to the node(s) it occupies—that have unique labels from its own perspective. Contracted particles have six ports while expanded particles have ten (see Fig. 3c). The particles are assumed to have a common sense of clockwise direction (a.k.a. chirality), but do not share a coordinate system or global compass. Thus, particles can label their ports in clockwise order starting from a local direction 0, but may have different *orientations* in $O = \{0, 1, ..., 5\}$ encoding their offsets for local direction 0 from global direction 0 (to the right).

For example, in Fig. 3c, the particle on the right has orientation 0 (i.e., it agrees with the global compass) while the particle on the left has orientation 4 (i.e., its local direction 0 is global direction 4). When a particle expands, it keeps its port labeling consistent by assigning label 0 to a port facing local direction 0 and then labeling the remaining ports in clockwise order.[1] In this way, it can recover its original labeling when it later contracts. A particle p communicates

---

[1] Note that there may be ambiguity in choosing a port facing local direction 0; e.g., in Fig. 3c, both port 0 and port 8 face local direction 0. In this case, the port facing local direction 0 and "away" from the particle is labeled 0.

with a neighbor q by placing a flag from the constant-size alphabet Σ on its port facing q. This can be thought of as p sending a message for q to read when q is next activated. Conversely, p receives information from q by reading the flag q has placed on its port facing p. The flag alphabet Σ is assumed to contain the "empty flag" ϵ to be used when no information is being communicated.

Particles move via a series of *expansions* and *contractions*: a contracted particle can expand into an unoccupied adjacent node to become expanded, and may then contract to occupy a single node once again. An expanded particle's *head* is the node it last expanded into and the other node it occupies is its *tail*; a contracted particle's head and tail are the same. If an expanded particle contracts into its head node, it has moved. Otherwise, contracting back into its tail node can be thought of as the particle exploring potential location to which it could expand but deciding not to over the course of two activations. Neighboring particles can coordinate their movements in a *handover*, which can occur one of two ways. A contracted particle p can "push" an expanded neighbor q by expanding into one of the nodes occupied by q, forcing q to contract. Alternatively, an expanded particle q can "pull" a contracted neighbor p by contracting, forcing p to expand into the node it is vacating. During its movements, each particle maintains a *tail port* in $T = \{0, 1, \ldots, 9\} \cup \{\varepsilon\}$ denoting the port furthest from its head if it is expanded or ε if it is contracted (see Fig. 3c). This information serves as the particle's memory about whether or not it is expanded, and, if so, what direction its tail is relative to its head.

More formally, the set of all possible movements is $M = \{\text{idle}\} \cup \{\text{expand}_i : i \in 0, 1, \ldots, 5\} \cup \{\text{contract}_i : i \in 0, 1, ..., 9\} \cup \{\text{handover}_i : i \in 0, 1, ..., 5\}$. An idle move simply means the particle does not move. If a particle p performs expand_i, p expands into the node its i-th port faces only if p is contracted and that node is unoccupied. If a particle p performs contract_i, p contracts out of the node incident to its i-th port only if p is expanded. The handover_i moves are not push or pull handover specific, nor do they actually perform the handover movements described above. Instead, a particle p performs handover_i when it initiates a handover with the neighbor its i-th port faces, say q. This initiation only succeeds if a neighboring particle q actually exists and p is contracted while q is expanded (or vice versa). To aid in executing the initiated handover— which will be described shortly—each particle keeps an *expansion direction* in $E = 0, 1, ..., 5 \cup \{\epsilon\}$ denoting the local direction it would like to expand in or ϵ if no expansion is needed.

The amoebot model assumes that particle systems progress by individual particles performing atomic actions asynchronously, where each particle independently and continuously executes its own instance of the given algorithm at potentially varying speeds. Assuming any conflicts that may arise in this concurrent execution are resolved—as is the case in the amoebot model, see [7]—a classical result under the asynchronous model states that there is a sequential ordering of atomic actions producing the same end result. Thus, we assume there is an *activation scheduler* responsible for activating exactly one particle at a time. This scheduler is assumed to be *fair*: each particle is assumed to be activated infinitely often. When a particle p is activated by the scheduler, it computes its *transition function* δ and applies the results:

$$\delta : Q \times \Sigma^{10} \times T \times E \to \mathcal{P}(Q \times \Sigma^{10} \times T \times E \times M).$$

For a given algorithm under the amoebot model, Q is a constant-size set of particle *states* while the flag alphabet Σ, the tail ports T, the expansion directions E, and the movements M are as defined above. The transition function δ allows a particle to use its state, its neighbors' flags facing it, its tail port, and its expansion direction to update these values and decide whether to move. If δ maps a unique input to multiple outputs, one output set is chosen arbitrarily. δ largely depends on the algorithm being executed; here, we describe a few general rules for δ our simulation will consider. Suppose that $\delta(q, (f_0, f_1, \ldots, f_9), t, e) = (q', (f'_0, f'_1, \ldots, f'_9), t', e', m')$.

- The movement m' must be valid according to the defined movement rules; e.g., if $m' = \mathsf{expand}_i$, particle p must be contracted and the node its i-th port faces must be unoccupied.
- If $t = \varepsilon$, then $f_6 = \cdots = f_9 = \epsilon$; i.e., if particle p is contracted, it cannot set flags for ports it doesn't have. This holds also for t' and $(f'_6, \ldots, f'_9)$.
- If $e \neq \epsilon$, then $t = \varepsilon$; i.e., particle p can only intend to expand in local direction e if it is contracted. This holds also for e' and t'.
- If $m' = \mathsf{expand}_i$, then $t' \neq \varepsilon$ and $e' = \epsilon$; i.e., if particle p expands in local direction i, it will be expanded (setting t' to the label opposite i after expansion) and should not intend to expand again immediately.
- If $e \neq \epsilon$, then either $m' = \mathsf{expand}_e$ or $m' = \mathsf{idle}$. That is, if particle p intends to expand in local direction e this activation, it either does so or has to wait.
- If $m' = \mathsf{contract}_i$, then $t' = \varepsilon$.

It remains to describe how handovers are executed with respect to δ. In a concurrent execution, a handover is performed as a coordinated, simultaneous expansion and contraction of two neighboring particles. In our sequential setting, however, we instead use a local synchronization mechanism to ensure the contracting particle moves first, followed by the expanding particle. To achieve this, we make one change to the scheduler. Whenever a particle p returns a movement $m' = \mathsf{handover}_i$ as output from δ, the scheduler finds the neighbor q facing the i-th port of p and ensures that the next three particles to be activated are q, then p, then q again.[2] We first describe a pull handover initiated by an expanded particle p with a contracted neighbor q.

1. Suppose p is chosen by the scheduler. Based on its state, its neighbors' flags facing it, its tail port indicating it is expanded, and its (empty) expansion direction, suppose δ returns $m' = \mathsf{handover}_i$. δ must also set f'_i to a *handover flag* indicating that p has initiated a handover with its neighbor.

---

[2] Note that this forced scheduling is simply a result of our formalism and does not alter or subvert the underlying asynchrony assumed by the amoebot model.

2. On seeing $m' = $ handover$_i$ returned, the scheduler finds neighbor q (the neighbor faced by the i-th port of p) and schedules $[q, p, q]$ as the next three particles to be activated. It activates q.

3. Based on the inputs to δ for particle q, and in particular the handover flag f_j from p and the fact that it is contracted, δ must evaluate such that q sets f'_j as a *will-expand flag*, sets $e' = j$, and sets $m' = $ idle.

4. The scheduler now has $[p, q]$, so it activates p.

5. Based on the inputs to δ for particle p, and in particular the will-expand flag f_i from q, δ must evaluate such that it clears $f'_i = \epsilon$ and sets $m' = $ contract$_i$ (setting $t' = \varepsilon$, following the rules above). Thus, it contracts.

6. The scheduler now has $[q]$, so it activates q.

7. Based on the inputs to δ for particle q, and in particular its expansion direction $e = j$, δ must evaluate such that q clears $f'_j = \epsilon$ and sets $m' = $ expand$_e$ (setting t' to the corresponding tail port opposite e).

8. The scheduler has no queued activations, so it chooses arbitrarily but fairly.

A push handover initiated by a contracted particle p with an expanded neighbor q is handled similarly. The first activation of p is the same as Step 1 above, causing the scheduler to do the same queuing as in Step 2. However, in Step 3, q sees the handover flag but also that it is expanded, meaning this is a push handover. Note, however, that a push handover is symmetric to a pull handover with the exception of which particle initiates; i.e., p performing a push handover with q yields the same result as q performing a pull handover with p. So, on seeing this is a push handover, q simply proceeds as particle p starting in Step 1, effectively exchanging roles with p.

The *configuration* of a particle p is $C(p) = (\boldsymbol{v}, o, q, t, e, (f_0, \dots, f_9))$, where $\boldsymbol{v} \in V$ is the coordinates of its head node, $o \in O$ is its orientation, $q \in Q$ is its state, $t \in T$ is its tail port, $e \in E$ is its expansion direction, and each $f_i \in \Sigma$ is the flag on its i-th port, for $i \in \{0, 1, \dots, 9\}$. Note that although the configuration of a particle p includes all information needed to reconstruct p, particle p itself does not have access to any global information or unique identifiers; in particular, it has no knowledge of $\boldsymbol{v}$ or o. The configuration of a particle system P is $C^*(P) = \{C(p) : p \in P\}$, the set of all configurations of particles in P. A system configuration is *valid* if no two particles in the system occupy a common node in G_Δ. We define $\mathcal{C}(P)$ to be the set of all valid system configurations of P. An *amoebot system* is defined as a 5-tuple $\mathcal{A} = (Q, \Sigma, \delta, P, \sigma)$, where Q is a constant-size set of particle states, Σ is a constant-size alphabet of flags, δ is the transition function, P is the particle system, and $\sigma \in \mathcal{C}(P)$ is the initial system configuration of $\mathcal{A}$ mapping each particle to its starting configuration.

For system configurations $\alpha, \alpha' \in \mathcal{C}(P)$, where $\alpha \neq \alpha'$, we say α yields α' (denoted $\alpha \rightarrow^\mathcal{A} \alpha'$) if α can become α' after a single particle activation. We use $\alpha \rightarrow^\mathcal{A}_* \alpha'$ if α yields α' in 0 or more activations. A sequence of configurations $(\alpha_0, \alpha_1, \dots, \alpha_k)$ is a *valid transition sequence* if for every $i \in [k]$ we have $\alpha_i \in \mathcal{C}(P)$, $\alpha_i \neq \alpha_{i+1}$, and $\alpha_i \rightarrow^\mathcal{A} \alpha_{i+1}$. A configuration $\alpha \in \mathcal{C}(P)$ is called *reachable* if there exists a valid transition sequence beginning at the initial configuration

σ and ending at α. A configuration $\alpha \in \mathcal{C}(P)$ is called *terminal* if there is no configuration $\alpha' \in \mathcal{C}(P)$ such that $\alpha \to^{\mathcal{A}} \alpha'$. A set of configurations $\Gamma \subseteq \mathcal{C}(P)$ is called *terminal* if for all $\alpha \in \Gamma$ there is no configuration $\alpha' \notin \Gamma$ such that $\alpha \to^{\mathcal{A}} \alpha'$ (i.e., no configuration in Γ can transition to any configuration outside of Γ). An amoebot system $\mathcal{A} = (Q, \Sigma, \delta, P, \sigma)$ is called *directed* if every transition sequence from σ leads to the same terminal configuration, or *directed to set Γ* if every transition sequence from σ leads to a configuration in Γ. Finally, given a shape S (i.e., a connected set of nodes in G_Δ), we say that system $\mathcal{A}$ *forms shape S* if and only if, for some set of configurations $\Gamma \subseteq \mathcal{C}(P)$, $\mathcal{A}$ is directed to Γ and for every $\alpha \in \Gamma$, the locations of the particles in α are exactly the locations of S (up to translation and rotation).

4   Simulating Amoebot Systems with Tile Automata

In this section, we present our main result, which is a construction that takes as input an amoebot system and which outputs a Tile Automata system that simulates it. However, we must first define what we mean by the term "simulate" in this context.

4.1   Defining Simulation

Intuitively, our simulation of an amoebot system by a Tile Automata system will consist of groups of tiles, called *macrotiles*, which each represent a single amoebot particle. Starting from an assembly which maps (via a mapping function to be described) to the initial configuration of an amoebot system, singleton tiles as well as macrotiles will attach, detach, and change states. Any changes to the assembly, modulo a scale factor, will map to new, valid configurations of the amoebot system. Conversely, for any valid configuration change of the amoebot system, the assembly will be able to change in such a way that it represents the new amoebot configuration, under the mapping function.

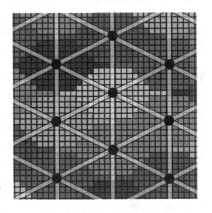

Fig. 4. A portion of the tessellation by the macrotile shape of our construction (shown in Figs. 6 and 5) with an overlay of G_Δ.

A *macrotile* is a connected, finite region of the plane $\mathbb{Z}^2$, whose shape can be any polyomino composed of connected unit squares. For a macrotile shape M to be valid to use for a simulation, it must tessellate. Since we are defining simulation of amoebot systems, which are embedded in the triangular grid, by Tile Automata, which are embedded in the square grid, a further condition is required for macrotile shapes. Let $\mathbb{T}(M)$ be a tessellation of the plane by macrotiles of shape M. Let G be the graph formed where every node is a macrotile in $\mathbb{T}(M)$ and there is an edge between a pair of nodes if and only if they are adjacent to each other in $\mathbb{T}(M)$. Then, graph G

must be isomorphic to the triangular grid graph (i.e. the graph of the triangular grid where each intersection is a node). This means each macrotile has the same 6-neighbor neighborhood as nodes in the triangular grid graph (see Fig. 4).

Let v be the coordinates of a node in G_Δ, and let m_v be the macrotile location which corresponds to it. Given Tile Automata system Γ and its set of producible assemblies PROD_Γ, for assembly $\mathcal{A} \in \text{PROD}_\Gamma$, let $\alpha \in A$ be a positioned assembly of A, and let $\alpha|m_v$ be the (possibly empty) subassembly of α contained in the locations of m_v. Given Γ and an amoebot system $\mathcal{A} = (Q, \Sigma, \delta, P, \sigma)$, a *macrotile representation function* R, from Γ to $\mathcal{A}$, is a function which takes as input the portion of an assembly contained within a single macrotile locations, and which returns either information about the configuration of an amoebot particle from P, or ϵ (which maps to empty space). That is, given some $\alpha|m_v$, $R(\alpha|m_v) \in \{(t, e, o, q, (f_0, f_1, ..., f_9)) \mid t \in T, o \in O, q \in Q, e \in E,$ and $f_i \in \Sigma\} \cup \{\epsilon\}$, where $t \in T$ is the relative direction of a particle's tail from its head, $o \in O$ is its orientation offset, $q \in Q$ is its state, and each $f_i \in \Sigma$, for $0 \leq i < 9$, is the flag in its ith port. An *assembly representation function*, or simply *representation function*, R^* from Γ to $\mathcal{A}$ takes as input an entire positioned assembly of Γ and applies R to every macrotile location and returns a corresponding amoebot system configuration from $\mathcal{C}(P)$.

For a positioned assembly $\alpha \in \text{PROD}_\Gamma$ such that $R^*(\alpha) = \alpha' \in \mathcal{C}(P)$, α is said to map *cleanly* to α' under R^* if for all non empty blocks $\alpha|m_v \in$ dom α, $v \in$ dom α' or $v' \in$ dom α' for some $v' = v + u$ where $u \in \{(1,0), (0,1), (-1,0), (0,-1), (-1,1), (1,-1)\}$. In other words, α may have tiles in a macrotile location representing a particle in α', or empty space in α' but only if that position is adjacent to a particle in α'. We call such growth "around the edges" of α *fuzz* and thus restrict it to be adjacent to macrotiles representing particles.

Note that the following definitions of *follows*, *models*, and *simulates*, as well as the previous definitions of macrotiles, fuzz, etc. are based upon similar definitions used to prove results about simulation and intrinsic universality in [8,12,18,19] and several other papers.

Definition 1 ($\mathcal{A}$ follows Γ). *Given Tile Automata system Γ, amoebot system $\mathcal{A}$, and assembly representation function R^* from Γ to $\mathcal{A}$, we say that $\mathcal{A}$ follows Γ (under R), and we write $\mathcal{A} \dashv_R \Gamma$, if $\alpha \rightarrow^\Gamma \beta$, for $\alpha, \beta \in \text{PROD}_\Gamma$, implies that $R^*(\alpha) \rightarrow^\mathcal{A}_* R^*(\beta)$.*

Definition 2 (Γ models $\mathcal{A}$). *Given Tile Automata system Γ, amoebot system $\mathcal{A} = (Q, \Sigma, \delta, P, \sigma)$, and assembly representation function R^* from Γ to $\mathcal{A}$, we say that Γ models $\mathcal{A}$ (under R), and we write $\Gamma \models_R \mathcal{A}$, if for every $\alpha \in \mathcal{C}(P)$, there exists $\Psi \subset \text{PROD}_\Gamma$ where $R^*(\alpha') = \alpha$ for all $\alpha' \in \Psi$, such that, for every $\beta \in \mathcal{C}(P)$ where $\alpha \rightarrow^\mathcal{A} \beta$, (1) for every $\alpha' \in \Psi$ there exists $\beta' \in \text{PROD}_\Gamma$ where $R^*(\beta') = \beta$ and $\alpha' \rightarrow^\Gamma \beta'$, and (2) for every $\alpha'' \in \text{PROD}_\Gamma$ where $\alpha'' \rightarrow^\Gamma \beta'$, $\beta' \in \text{PROD}_\Gamma$, $R^*(\alpha'') = \alpha$, and $R^*(\beta') = \beta$, there exists $\alpha' \in \Psi$ such that $\alpha' \rightarrow^\Gamma \alpha''$.*

Definition 2 essentially specifies that every time Γ simulates an amoebot configuration $\alpha \in \mathcal{C}(P)$, there must be at least one valid growth path in Γ for each of the possible next configurations that α could transition into from α, which results in an assembly in Γ that maps to that next step.

Definition 3 (Γ **simulates** $\mathcal{A}$). *Given Tile Automata system Γ, amoebot system $\mathcal{A}$, and assembly representation function R^* from Γ to $\mathcal{A}$, if $\mathcal{A} \dashv_R \Gamma$ and $\Gamma \models_R \mathcal{A}$, we say that Γ simulates $\mathcal{A}$ under R.*

With the definition of what it means for a Tile Automata system to simulate an amoebot system, we can now state our main result.

Theorem 1. *Let $\mathcal{A}$ be an arbitrary amoebot system. There exists a Tile Automata system Γ and assembly representation function R^* from Γ to $\mathcal{A}$ such that Γ simulates $\mathcal{A}$ under R. Furthermore, the simulation is at scale factor 100.*

To prove Theorem 1, we let $\mathcal{A} = (Q, \Sigma_{\mathcal{A}}, \delta, P, \sigma)$ be an arbitrary amoebot system. We will now show how to construct a Tile Automata system $\Gamma = (\Sigma_\Gamma, \Lambda, \Pi, \Delta, \tau)$ such that Γ simulates $\mathcal{A}$ at scale factor 100. The rest of this section contains details of our construction.

4.2   Construction Definitions

Neighborhood - In the geometric amoebots model, particles are aware of the occupation of all locations on the lattice adjacent to their own. The neighborhood of a given location on the lattice is the set of its six neighbors. Pertaining to a particle, we say that a particle's neighborhood is the set all particles occupying adjacent locations on the lattice, defined by $N(p)$, where p is a particle. Note that $|N(p)| \le 6$ if p is contracted and 10 if it is expanded.

Macrotile - A τ-stable assembly of TA tiles g such that the macrotile representation function R maps g to a valid particle in $\mathcal{A}$. This simulation makes use of macrotiles with an approximately hexagonal shape and special tiles within each macrotile used to calculate information about its movement and neighborhood. See Fig. 5 for an overview.

Clock Tiles - The tiles at the middle of every particle macrotile used to keep track of state, flags, T value, and neighborhood information. The middle clock tile is responsible for maintaining the particle's state $q \in Q$, and the surrounding clock tiles (called subordinate clock tiles) combine information from the particle edges and neighbors to pass into the central clock tile.

Wire Tiles - Rows of tiles leading from the bank of clock tiles in the middle of every macrotile to each edge, purposed with transmitting information from the clock to the neighboring tiles and available edges.

Filler Tiles - Tiles that serve no function within a macrotile other than to maintain connectivity with other components and shape the macrotile.

Flag Tiles - Exposed wire ends on each side of the tile responsible for maintaining flag states from Σ in $\mathcal{A}$, as well as reading flags from their respective neighbors. Neighboring flags are retrieved via wire transmission.

Timing Tiles - Individual tiles in Γ that diffuse into specific slots in the particle macrotiles that "start" that particle's turn, and disconnect after the turn is finished. Timing tiles are inert except for connecting to a central clock tile, and serve as the "asynchronous clock" for our simulation.

ϵ ***Tiles*** - Individual tiles in Γ that attach to the available flag tiles of macrotiles with non-full neighborhoods who are querying or attempting to lock their neighborhood flags. ϵ tiles can only attach to the exposed end of a wire displaying a lock or query signal flag. After attaching, they serve only to undergo a single state transition, which indicates to the wire end that there exists no neighbor there. After this transition, the wire propagates this information back to its clock bank and the ϵ tile detaches.

Floating Macrotile - These macrotiles ("floats"), will represent (portions of) particles in $\mathcal{A}$ but are not connected to the swarm. They attach to valid sites along the perimeter and simulate the "head" of an expanding particle.

Configuration Tile - After macrotiles complete their turn, they combine all values of their configuration of which they are aware ($q \in Q, e \in E, (f_0, f_1, ..., f_9 \in \Sigma^{10}), t \in T$) into a single tile proximal to the clock bank called the *Configuration Tile*. When a macrotile is engaging in an expansion or handover move, the configuration tile is used by the representation function R to map the active macrotile to its previous configuration, until the transition completes. An active macrotile engaged in these moves will be mapped by its configuration tile until it's no longer displaying a flag indicating it is engaged in either an expansion or handover on any of its wires, ensuring the simulation of an atomic move. (For a more complete explanation, see appendix B of [1]s.)

4.3 Simulation Overview

The simulation of $\mathcal{A}$ by Γ is affected with the assistance of the hexagonal macrotiles and the signals implemented via TA state transitions. In [5], since there were only four directions from which signals could come, it was sufficient for each macrotile to have one clock tile, which would transition its own state based on signals received from wires and send its state down the wires. Since the geometric amoebots model exists on the G_Δ, signals can come from up to six directions, necessitating the use of multiple clock tiles. Figure 5 contains a high-level depiction of a macrotile in Γ that simulates a particle of $\mathcal{A}$.

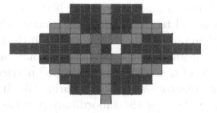

Fig. 5. Blue tiles are subordinate clock tiles, dark blue is the central clock tile, green tiles wires, and grey tiles filler. The empty location immediately east of the central clock tile is where the timing tile attaches to signal the central clock tile to begin the turn. Flags are displayed on the the outermost tile of every wire. (Color figure online)

Figure 6 illustrates a simple example of simulated particle movement. Macrotiles must be initially arranged into a configuration α that under $R^*(\alpha)$ maps to a valid configuration $\alpha' \in \mathcal{C}(P)$, with connected edges representing adjacency in α'. Macrotiles start with their respective states, flags, and t values set to whatever those states are for the corresponding particle in α'. Swarm macrotiles may then begin to accept timing tiles, starting their turns. We use neighborhood *lock signals* to ensure that no particles that are in the same neighborhood attempt to move at the same time, avoiding asynchronous conflicts. Expansion is facilitated by the attachment along perimeter sites of floating macrotiles. The authors additionally considered systems where moves progressed by growing a new macrotile wherever a particle wanted to expand, but this construction technique requires a longer wait between neighborhood locks and unlocks. In the interest of minimizing the overhead that simulation requires, we wanted to minimize the amount of time that a neighborhood had to be locked in order to encourage collaborative movement, and decided to use prefabricated floating macrotiles.

Once a particle macrotile has received a timing tile, it only continues its turn if it is not already locked by a neighbor. If the particle is not locked down, then the active macrotile sends signals to all of its neighbors to lock down its neighborhood and it can continue without fear of causing conflict. Should two *lock* signals be traveling towards each other along a shared wire between two macrotiles, whichever signal is first carried into the other macrotile's wires via state transitions overwrites the signal originating from the slower macrotile, and the faster propagating signal's originator locks down the slower. The neighboring flags are needed to simulate the transition function, and are sent from neighbors to the active macrotile via wire transmission. Once the active macrotile has decided its new state, flags, and move, it updates this information and attempts to execute its chosen move. If the move is a simple expansion, it marks the site where it wants to expand with a valid attachment signal and keeps the neighborhood locked until a floating macrotile connects to it, representing the "head" of the expanding particle. If the particle chooses to contract, it sends signals to the tail to detach from the swarm, whereupon it will become another float, and then unlocks its neighborhood. If the particle chose a handover$_i$, it performs some additional checks to ensure viability and then sends signals either "taking over" one of a neighboring expanded particle's macrotiles, or ceding control of one of its macrotiles to a neighboring tile. In the case of handover$_i$, an additional state and flags for the subordinate particle are returned from the transition function, to be propagated from the active macrotile to the subordinate macrotile via wire transmissions as control of the macrotile changes hands.

5   Simulation of Movement

Before the simulation begins, given $\mathcal{A}$ with initial configuration σ, we produce Λ for Γ which consists of a connected configuration of macrotiles, each macrotile mapping to a corresponding particle in σ under R. To capture the asynchronous nature of amoebot particle activations, we utilize timing tile diffusion into macrotiles to "activate" them for their turns. After attachment, the tile

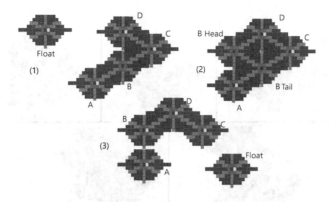

Fig. 6. Particle B simulates movement by allowing a float to attach to a free edge, then detaching the original macrotile representing particle B over the course of two activations.

sends a signal to the central clock to start its turn. After a given macrotile has started its turn and has successfully locked its neighborhood, it gathers all information necessary for its transition function via wire-propagated signals (detailed in the technical appendix). To ensure that the transition function affected by macrotiles is isomorphic to the transition function affected by $\mathcal{A}$, we combine all of the flags, e and t values (resulting in a $|\Sigma^{10}| * |t| * |e|$ increase in state complexity for clock tiles) into the tile to the left of the central clock tile. Once this value is at that tile, the central clock tile, which holds the state, undergoes a state transition defined in the construction of Δ.

Once the new state, flags, and move are produced, the particle propagates its flags down their respective wires and in the case that handover$_i$ is returned for the value of m, the particle additionally sends a signal (detailed in the technical appendix) to its ith neighbor to ensure that the neighbor has the proper orientation to facilitate that move. For expand$_i$, the active macrotile sends a signal down the wire in the ith direction that allows a float to attach to that edge. After connection of a float, the active macrotile further sends a *CopySignal* to the newly attached float so the new float can copy the states and relevant flags of the expanding macrotile and fully become the head. The float sends an *AcknowledgementSignal* after it is displaying the proper state and flags, which tells the newly expanded macrotile that it's safe to unlock its neighborhood. For contract$_i$, the expanded macrotile sends a *Detach* signal to the macrotile that contains the ith port. After detachment, the recently contracted macrotile unlocks its neighborhood. For idle, states and flags may be updated, but no change to the orientation of the simulated particles occurs. Once a macrotile executing an idle move changes its state and sends the new flags to its flag tiles, it unlocks its neighborhood and ends its turn. Figure 7 depicts the series of steps of a move containing a handover, which is described in the following section.

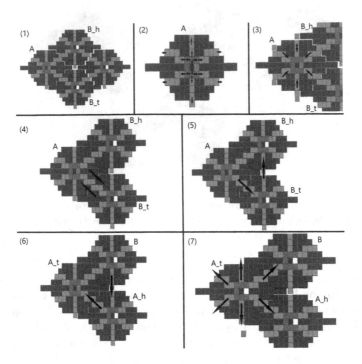

Fig. 7. (1) The initial configuration of a subset of a valid swarm. Note that only A has a purple timing tile to the right of its Central Clock tile. (2) The flow of the $LockSignal_i$ sent out from A's CC tile. (3) The responses from A's wires flow back to the CC tile. Unoccupied edges allow for the attachment of yellow epsilon tiles, which indicate to a given wire that it has no macrotile neighbor. A has no neighbors attempting their own move, and is safe to continue its move. (4) A is attempting to execute a handover, and so must check to ensure that the subordinate particle it intends to use is in the proper configuration. It sends out a $handoverConfirm_i$ signal, to which B responds with an $Acknowledgement$ signal containing B's t value. (5) Since B is in the proper configuration to enable a $handoverPush_i$, A continues by sending a $handoverPush_i$ signal to the subordinate macrotile it wants to take over. The subordinate macrotile B_t strips the flags, state, and new t value pertinent to itself and propagates the remaining new flags, state and t value to its other half B_h. (6) B_h alters its state and flags in response to the signal it has just received and sends back an acknowledgement signal to its tail. As this signal travels through the shared wires between B_h and B_t, it clears the $handover$ flags which indicate to the representation function that it should check those macrotiles' configuration tile for information instead of flags and the CC tile for the state. The clearing of these flags is the event that precipitates the transition of the representation of the macrotile displaying them, so that $R^*(\alpha) \neq R^*(\beta)$, where α and β are the macrotile before and after this step. (7) Finally, A_t receives the acknowledgement signal propagated from what now maps to A_h, so it sends out a final $Unlock$ signal to all of its neighborhood and A ends the turn by the CC undergoing a transition with the timing tile, rendering it inert and without affinities, thereby detaching it. (Color figure online)

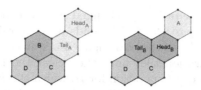

Fig. 8. A *handoverPull_i* movement executed by particle A. The overall orientation of the swarm does not change, but $Tail_A$ changes hands and becomes the head of B, the subordinate particle in this exchange.

5.1 handover

The Amoebots model defines a move called the handover$_i$, which allows two neighboring particles, one expanded and one contracted, to simultaneously contract from and expand into the same grid location. This move can be initiated by either particle, and involves the initiating particle's neighbor in the *ith* direction. There are four possibilities for handover contracts: A contracted particle can expand into a spot previously occupied by a head, a contracted particle can expand into a spot previously occupied by a tail, an expanded particle can contract and "force" its neighbor into expanding into wherever its head is currently located, and an expanded particle can contract and "force" its neighbor into expanding into wherever its tail is currently located. It's necessary that both moves happen simultaneously to enable certain actions such as moving a collection of particles through a static tunnel of width one. For any *handoverPull* movement, the initiating particle must be expanded and the subordinate particle contracted, while for any *handoverPush* movement, the reverse is true (Fig. 8).

All moves involve locking down the active particle's neighborhood. Since handover$_i$ necessarily changes the orientation of both the initiating and subordinate particles, it is necessary to lock down the respective neighborhoods of all particles involved. Thus, the *handoverPull_i* movement requires a *ProgressiveLockdownSignal_i* to be sent to the subordinate particle(s) to be further propagated to their respective neighborhoods. Note, that for the duration of the execution of a handover move, a handover$_i$ flag will be displayed, ensuring that the macrotile maps to its previous configuration under the representation function R until the handover$_i$ or expand$_i$ flags are cleared from the macrotiles' respective ports. This ensures that we have smooth transitions for macrotiles mapping to the atomic transitions of amoebots.

Handover moves in our simulation occur after the initiating particle has locked down its own neighborhood, checked to ensure move validity with the handover$_i$ signal, and further sent a *ProgressiveLockdownSignal_i* to each subordinate particle. The initiating particle ensures that the configuration of expansions and contractions are appropriate for the move it is attempting, and then sends a signal to the macrotile that will be passed from one particle to another. This signal contains the new state and flags for the subordinate macrotile(s). Once these fields are updated, and the handover$_i$ flags are cleared from the macrotiles' wires, the macrotile that changed state is considered to be the head of the newly expanded particle, which sends an acknowledgement signal back to the initiating macrotile. Macrotiles engaged in a handover$_i$ move are mapped by R via their configuration tiles until they clear the flags, after which they revert to the normal mode of operation for R, which checks a macrotile's central clock tile and the subordinate clock tile to the immediate west. After receiving the acknowledgement of successful transition, the initiating macrotile sends out unlock signals to its neighborhood (which necessarily propagate from subordinate particles to their respective neighborhoods as well), and finally the initiating macrotile detaches its timing tile, ending the turn.

5.2   Attachment Sites

To avoid potential conflicts between tiles undergoing their state transitions and the floating macrotiles, we only allow the floating macrotiles to attach to valid attachment sites along the perimeter of the swarm. The perimeter of the swarm shares no affinity with the float tiles by default. Only after a perimeter particle undergoes its transition function and returns $M = \text{expand}_i$ and marks that edge with a state that has affinity with the float can any float attach to the particle that wants to expand. Floats attach to the swarm via a τ-strength attachment along the exposed wire end of the perimeter particle that is attempting to expand. The only time macrotiles can detach from the swarm is when a given expanded macrotile receives a *Detach* signal from its other end. When this occurs, the detaching macrotile sends signals from its clock bank to all edges to have them change their exposed wire ends to a state with no affinity to the swarm. Once the macrotile is no longer part of the swarm, it sends *query* signals to all of its edges. In the case that a macrotile receives six ϵ responses from its wires (that is, it has no neighbors) after a query signal, it undergoes states transitions in the clock and wires that make it a float. The newly contracted macrotile prevents unintentional reattachments of other floats because the *Detach* signal, after leaving the contracted particle's wire, leaves the wire in its respective flag state with no affinity with floats. The wire that corresponds to the detached macrotile will not allow attachments again until it receives an *Attach* signal from the central clock bank again.

6   Conclusion

We have presented a simulation construction in which an amoebots system can be simulated by a collection of Tile Automata macrotiles. The mechanisms by which particle movement is simulated were discussed as well, such as how the atomic actions of the amoebots model were replicated within the simulation without threat of interruption via state transitions. We hope this fits into a larger schema of comparing the power of various computational models by simulation.

Acknowledgements. The authors would like to thank Schloss Dagstuhl – Leibniz Center for Informatics and the organizers and participants of Dagstuhl Seminar 18331

"Algorithmic Foundations of Programmable Matter" [3]. The initial brainstorming and work for this paper began during that workshop and was inspired by many interesting discussions with the participants.

References

1. Alumbaugh, J.C., Daymude, J.J., Demaine, E.D., Patitz, M.J., Richa, A.W.: Simulation of programmable matter systems using active tile-based self-assembly. Technical Reports 1906.01773, Computing Research Repository (2019). http://arxiv.org/abs/1906.01773
2. Andrés Arroyo, M., Cannon, S., Daymude, J.J., Randall, D., Richa, A.W.: A stochastic approach to shortcut bridging in programmable matter. Nat. Comput. **17**(4), 723–741 (2018)
3. Berman, S., Fekete, S.P., Patitz, M.J., Scheideler, C.: Algorithmic foundations of programmable matter (dagstuhl seminar 18331). Dagstuhl Rep. **8**(8), 48–66 (2019). https://doi.org/10.4230/DagRep.8.8.48. http://drops.dagstuhl.de/opus/volltexte/2019/10235
4. Cannon, S., Daymude, J.J., Randall, D., Richa, A.W.: A Markov chain algorithm for compression in self-organizing particle systems. In: Proceedings of the 2016 ACM Symposium on Principles of Distributed Computing, pp. 279–288, PODC 2016. ACM, New York (2016)
5. Chalk, C., Luchsinger, A., Martinez, E., Schweller, R., Winslow, A., Wylie, T.: Freezing simulates non-freezing tile automata. In: Doty, D., Dietz, H. (eds.) DNA 2018. LNCS, vol. 11145, pp. 155–172. Springer, Cham (2018). https://doi.org/10.1007/978-3-030-00030-1_10
6. Daymude, J.J., Gmyr, R., Richa, A.W., Scheideler, C., Strothmann, T.: Improved leader election for self-organizing programmable matter. In: Fernández Anta, A., Jurdzinski, T., Mosteiro, M.A., Zhang, Y. (eds.) ALGOSENSORS 2017. LNCS, vol. 10718, pp. 127–140. Springer, Cham (2017). https://doi.org/10.1007/978-3-319-72751-6_10
7. Daymude, J.J., Hinnenthal, K., Richa, A.W., Scheideler, C.: Computing by programmable particles. In: Flocchini, P., Prencipe, G., Santoro, N. (eds.) Distributed Computing by Mobile Entities: Current Research in Moving and Computing. LNCS, vol. 11340, pp. 615–681. Springer, Cham (2019). https://doi.org/10.1007/978-3-030-11072-7_22
8. Demaine, E.D., Patitz, M.J., Rogers, T.A., Schweller, R.T., Summers, S.M., Woods, D.: The two-handed tile assembly model is not intrinsically universal. In: Fomin, F.V., Freivalds, R., Kwiatkowska, M., Peleg, D. (eds.) ICALP 2013. LNCS, vol. 7965, pp. 400–412. Springer, Heidelberg (2013). https://doi.org/10.1007/978-3-642-39206-1_34
9. Derakhshandeh, Z., Gmyr, R., Richa, A.W., Scheideler, C., Strothmann, T.: Universal shape formation for programmable matter. In: Proceedings of the 28th ACM Symposium on Parallelism in Algorithms and Architectures SPAA, pp. 289–299. 2016. ACM, New York (2016)
10. Derakhshandeh, Z., Gmyr, R., Richa, A.W., Scheideler, C., Strothmann, T.: Universal coating for programmable matter. Theor. Comput. Sci. **671**, 56–68 (2017)
11. Derakhshandeh, Z., Richa, A., Dolev, S., Scheideler, C., Gmyr, R., Strothmann, T.: Brief announcement: amoebot-a new model for programmable matter. In: Annual ACM Symposium on Parallelism in Algorithms and Architectures, pp. 220–222. Association for Computing Machinery (2014)

12. Doty, D., Lutz, J.H., Patitz, M.J., Schweller, R.T., Summers, S.M., Woods, D.: The tile assembly model is intrinsically universal. In: 2012 Proceedings of the 53rd Annual IEEE Symposium on Foundations of Computer Science, pp. 302–310. FOCS (2012)
13. Fekete, S.P., Hendricks, J., Patitz, M.J., Rogers, T.A., Schweller, R.T.: Universal computation with arbitrary polyomino tiles in non-cooperative self-assembly. In: 2015 Proceedings of the Twenty-Sixth Annual ACM-SIAM Symposium on Discrete Algorithms (SODA 2015), San Diego, CA, USA, pp. 148–167, 4–6 January 2015
14. Gilbert, O., Hendricks, J., Patitz, M.J., Rogers, T.A.: Computing in continuous space with self-assembling polygonal tiles. In: 2016 Proceedings of the Twenty-Seventh Annual ACM-SIAM Symposium on Discrete Algorithms (SODA 2016), Arlington, VA, USA, pp. 937–956, 10–12 January 2016
15. Hendricks, J., Padilla, J.E., Patitz, M.J., Rogers, T.A.: Signal transmission across tile assemblies: 3D static tiles simulate active self-assembly by 2D signal-passing tiles. In: Soloveichik, D., Yurke, B. (eds.) DNA 2013. LNCS, vol. 8141, pp. 90–104. Springer, Cham (2013). https://doi.org/10.1007/978-3-319-01928-4_7
16. Hendricks, J., Patitz, M.J., Rogers, T.A., Summers, S.M.: The power of duples (in self-assembly): it's not so hip to be square. Theor. Comput. Sci. **743**, 148–166 (2015)
17. Kao, M.-Y., Schweller, R.: Randomized self-assembly for approximate shapes. In: Aceto, L., Damgård, I., Goldberg, L.A., Halldórsson, M.M., Ingólfsdóttir, A., Walukiewicz, I. (eds.) ICALP 2008. LNCS, vol. 5125, pp. 370–384. Springer, Heidelberg (2008). https://doi.org/10.1007/978-3-540-70575-8_31
18. Meunier, P.E., Patitz, M.J., Summers, S.M., Theyssier, G., Winslow, A., Woods, D.: Intrinsic universality in tile self-assembly requires cooperation. In: 2014 Proceedings of the ACM-SIAM Symposium on Discrete Algorithms (SODA 2014), Portland, OR, USA, pp. 752–771, 5–7 January 2014
19. Meunier, P., Woods, D.: The non-cooperative tile assembly model is not intrinsically universal or capable of bounded turing machine simulation. In: 2017 Proceedings of the 49th Annual ACM SIGACT Symposium on Theory of Computing, STOC 2017, Montreal, QC, Canada, pp. 328–341, 19–23 June 2017
20. Padilla, J.E., Patitz, M.J., Schweller, R.T., Seeman, N.C., Summers, S.M., Zhong, X.: Asynchronous signal passing for tile self-assembly: Fuel efficient computation and efficient assembly of shapes. Int. J. Found. Comput. Sci. **25**(4), 459–488 (2014)
21. Patitz, M.J., Schweller, R.T., Summers, S.M.: Exact shapes and turing universality at temperature 1 with a single negative glue. In: 2011 Proceedings of the 17th International Conference on DNA Computing and Molecular Programming, pp. 175–189. DNA (2011)
22. Winfree, E.: Algorithmic self-assembly of DNA. Ph.D. thesis, California Institute of Technology, June 1998
23. Woods, D.: Intrinsic universality and the computational power of self-assembly. Philos. Trans. R. Soc. Lond. A: Math. Phys. Eng. Sci. **373**(2046), 1–13 (2015)
24. Woods, D., Chen, H.L., Goodfriend, S., Dabby, N., Winfree, E., Yin, P.: Active self-assembly of algorithmic shapes and patterns in polylogarithmic time. In: Proceedings of the 4th Conference on Innovations in Theoretical Computer Science, ITCS 2013, pp. 353–354. ACM, New York (2013)

Combined Amplification and Molecular Classification for Gene Expression Diagnostics

Gokul Gowri[1], Randolph Lopez[2], and Georg Seelig[2(✉)]

[1] California Institute of Technology, Pasadena, CA, USA
[2] University of Washington, Seattle, WA, USA
gseelig@uw.edu

Abstract. RNA expression profiles contain information about the state of a cell and specific gene expression changes are often associated with disease. Classification of blood or similar samples based on RNA expression can thus be a powerful method for disease diagnosis. However, basing diagnostic decisions on RNA expression remains impractical for most clinical applications because it requires costly and slow gene expression profiling based on microarrays or next generation sequencing followed by often complex *in silico* analysis. DNA-based molecular classifiers that perform a computation over RNA inputs and summarize a diagnostic result *in situ* have been developed to address this issue, but lack the sensitivity required for use with actual biological samples. To address this limitation, we here propose a DNA-based classification system that takes advantage of PCR-based amplification for increased sensitivity. In our initial scheme, the importance of a transcript for a diagnostic decision is proportional to the number of molecular probes bound to that transcript. Although probe concentration is similar to that of the RNA input, subsequent amplification of the probes with PCR can dramatically increase the sensitivity of the assay. However, even slight biases in PCR efficiency can distort weight information encoded by the original probe set. To address this concern, we developed and mathematically analyzed multiple strategies for mitigating the bias associated with PCR-based amplification. We evaluate these amplified molecular classification strategies through simulation using two distinct gene expression data sets and associated disease categories as inputs. Through this analysis, we arrive at a novel molecular classifier framework that naturally accommodates PCR bias and also uses a smaller number of molecular probes than required in the initial, naive implementation.

1 Introduction

Detection and quantification of RNA molecules in blood or tissue can be a powerful tool for disease diagnosis. Although detection of a single, differentially expressed molecular marker might be ideal, such a distinctive marker may not exist for a disease state of interest. Instead, a diagnostic decision may have to be

© Springer Nature Switzerland AG 2019
C. Thachuk and Y. Liu (Eds.): DNA 25, LNCS 11648, pp. 159–173, 2019.
https://doi.org/10.1007/978-3-030-26807-7_9

based on panels of differentially expressed genes and may require careful weighing of the contributions of each gene in the panel. The conventional workflow for building such multi-gene classifiers consists of experimentally measuring gene expression in different "states" (e.g. healthy and disease) using microarrays or high-throughput sequencing, followed by training of a computational classifier that learns to assign labels to the samples based on differences in gene expression between those states. Once the classifier is trained it can be used to label previously unseen samples based on their gene expression profiles, thus providing a powerful tool for aiding diagnostics decisions. For example, an *in silico* whole blood gene expression classifier has been developed to distinguish bacterial infections, viral infections, and non-infectious disease [1]. Similarly, an *in silico* blood platelet gene expression classifier has been developed to distinguish six different types of cancer [2]. Both of these classifiers were trained using support vector machine (SVM) methods, and each involved more than 100 features (i.e. differentially expressed genes) [1,2]. These examples provide proof-of-principle for the power of a diagnostic approach based on analysing multi-gene panels but the complexity and cost of gene expression classification has limited its implementation to a research setting and to a small subset of clinical problems.

Using molecular computation to implement a disease classifier could overcome some of the limitations of existing technologies. In such an approach, the molecular diagnostic "computer" could be mixed with the patient sample, perform an analysis of the sample through a sequence of molecular interactions and then summarize the result into an easy-to-interpret signal that represents the diagnosis. If realized, such an approach could accelerate the diagnosis by simplifying the measurement process and eliminating the need for separate computational analysis.

DNA-based circuits have long been proposed as a platform for performing gene expression diagnostics [3]. In our own previous work, we demonstrated a platform to translate a linear classifier into a DNA strand displacement circuit that can operate on a set of RNA inputs and classify a sample by calculating a weighted sum over those inputs [4]. However, our work was limited to a proof-of-principle implementation with *in vitro* transcribed RNA since the concentration of RNA molecules in biological samples is significantly below the detection limit of that assay.

In this work, we propose a method for combining targeted RNA amplification and molecular classification for gene expression diagnostics. We begin by proposing a classifier design that combines computation with enzymatic amplification (Sect. 2). In our approach, molecular probes are used to assign weights to each RNA feature in the classifier and PCR-based amplification is used to amplify the signal resulting from each probe. The workflow for such an approach is shown in Fig. 1. We then demonstrate a computational approach, starting from gene expression data, for learning a sparsely-featured classifier that is well-suited for molecular implementation (Sect. 3). However, this formal framework assumes that molecular classifier components behave ideally, while in practice nucleic acid amplification mechanisms are biased, i.e. different probe sequences get amplified to varying degrees which distorts the classification result (Sect. 4). We then explore the effects of amplification bias and developed strategies to mitigate amplification bias and maintain classification accuracy. First, we explore a strategy for averaging over multiple probes, each with an unknown bias (Sect. 5). Although successful to an extent, we will argue that this naive strategy is impractical because of the large number of probes required. Moreover, because in practice the amplification bias is consistent across samples, it is possible to actually measure the probe bias and incorporate that information into the classifier design. Second, we thus ask whether accurate classification can be achieved if individual probes are biased but well-characterized (Sect. 6). We will show that this approach results in a more compact molecular implementation of the classifier but that classification accuracy is still limited compared to an ideal classifier. Finally, we develop an approach wherein each weight associated with a transcript is implemented by just two competing probes, one associated with a negative and the other with a positive weight (Sect. 7). We show that even with biased probes any target value of the weight can be closely approximated as long as we have precise control over probe concentrations. We validate each of our designs by simulating molecular classification of gene expression profiles in the context of cancer and infectious disease diagnostics. Our results indicate that probe characterization and subsequent selection enable the construction of sensitive, accurate and cost-effective molecular classifiers for gene expression diagnostics.

2   A Molecular Classifier with Built-In Amplification

Here, we lay out a strategy for creating a molecular classifier with a built-in amplification mechanism. We first review our earlier classifier design that provides that basis for our work and then discuss how an enzymatic amplification step can be used to dramatically increase the specificity of the earlier approach.

In the molecular classifiers reported in [4], the magnitude of a weight associated with a transcript is implemented by the number of probes designed to target that transcript; an identity domain appended to the hybridization domain indicates the sign of a each probe and thus of each weight. The total concentration of all positive (negative) identity domains thus represents the sum of all positive (negative) weights in the system. Identity domains interact with a downstream DNA circuit that performs subtraction through pairwise annihilation of positive and negative signal [4,5]. The remaining signal is converted to fluorescent output. Alternatively, if reporter constructs labeled with distinct fluorophores are used to separately read out the concentrations of positive and negative identity domains, the subtraction can be performed in silico. In either case, this approach requires weights to be small (<10) integers, because each unit weight requires a unique binding site on a transcript. This design is demonstrated in Fig. 2. Moreover, in this framework, the output signal has similar magnitude to the input signal, so sensitivity is low, and not sufficient for clinical disease diagnostics.

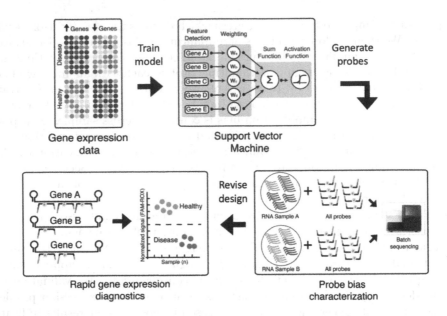

Fig. 1. Workflow for combined amplification and classification of RNA samples. An in-silico classifier is trained and validated on publicly available gene expression data. Then, multiple probes are designed and synthesized to target the set of genes in the classifier. This set of probes is amplified using an RNA sample with known gene expression data. Using next-generation sequencing, the number of amplified probes can be counted in batch in order to determine probe specific amplification bias. Subsequently, this data informs a design for molecular classifier that will implement the desired classification model.

Increasing Sensitivity. Amplification reactions based on strand displacement cascades [6–9] provide one interesting approach to signal amplification and such systems have even been used to create amplifiers with controllable gain that can be combined into linear classifier circuits [10,11]. Still, the gain that can be achieved with a single strand displacement-based catalytic amplifier is typically limited and cascades of multiple amplifiers that might have higher gain are often leaky or simply not robust enough for practical applications.

In order to address the lack of sensitivity intrinsic to our initial classification scheme we thus compared several enzyme-based approaches for targeted RNA amplification including NASBA [14], rolling-circle amplification [15,17,18], RASL [12] and multiplex PCR [19]. Upon evaluating and characterizing several of these methods, we decided to implement a molecular classification workflow using RASL probes. RASL (RNA-mediated oligonucleotide annealing, selection, and ligation) is commonly used for targeted RNA sequencing for the purpose of gene expression quantification [12].

Probe Amplification with RASL. In the RASL protocol, shown in Fig. 3a, custom probe pairs are designed for each RNA of interest. The two probes of

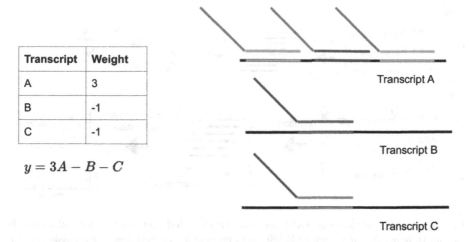

Transcript	Weight
A	3
B	-1
C	-1

$$y = 3A - B - C$$

Fig. 2. An example molecular implementation of a linear classifier with the approach previously demonstrated in [4]. The red and blue domains represent positive and negative values respectively for downstream subtraction and reporting mechanisms. (Color figure online)

a pair bind neighboring sites on an mRNA target such that they create a contiguous sequence with a nick at the center. Adjacent probes are then ligated together. Importantly, ligation can only occur if probes are bound to the RNA. Each probe also contains universal amplification overhangs. PCR amplification of the ligated probes enables accurate amplification of low total RNA amounts (10 nanograms) for gene expression quantification using next-generation sequencing. Because only ligated probes can be efficiently amplified and because ligation requires the RNA template, RASL is specific and quantitative method for RNA detection.

A molecular classifier using RASL probes is shown in Fig. 3b. Different integer weights can be implemented by varying the number of probes per transcript. To implement positive and negative weights, each probe also contains a sequence barcode that triggers either a green or red fluorescent reporter during amplification. The difference between the aggregate fluorescent signals from the two channels is the classification output.

Amplification by RASL is consistent across samples, but is inconsistent within samples, such that certain probes are amplified with greater propensity than others [12]. As a result, n probes will not necessarily yield n times as large a signal as a single probe representing unit weight, so molecular classifiers designed with the approach of [4] will not behave similarly to the *in silico* classifier upon amplification. Below we discuss how such bias can be accounted for in the classifier design.

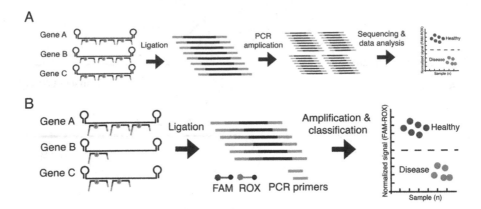

Fig. 3. Overview of existing RASL-seq and our modified method for molecular classification. In panel A, the standard RASL-seq protocol for targeted gene expression counting using next-generation sequencing is shown. Multiple probe pairs are hybridized to their RNA targets followed by ligation of gaps. The yellow circled P's indicate the 5' phosphates required for ligation. The ligated products can be amplified using common primers and then used as input for next generation sequencing. Sequencing data is then used for gene expression analysis. In panel B, the protocol for molecular classification using RASL is shown. Our modified RASL probes contain a positive or negative barcode (red or green domain) associated with each probe. The number of probes that bind to each transcript can be varied to tune the effective amplification weight on each transcript. During amplification, two fluorescent reporters (FAM and ROX) are triggered based on the presence of each barcode. This fluorescent signal can then determine the classification outcome for a given sample. (Color figure online)

3   Computational Design of Sparsely Featured Diagnostic Classifiers

Before building a molecular classifier for gene expression diagnostics, we first train an *in silico* disease classifier on available gene expression data. This computational classifier is then mapped onto a molecular classifier. In a molecular classifier, each feature corresponds to a target transcript, and probes are implemented as oligonucleotides that can bind to a unique region within a transcript. Although feasible in principle, molecular classifiers with large numbers of features are currently not practical because of the large numbers of probes required to implement them. As s first step in the classifier design workflow we thus aim to create sparsely featured yet clinically valuable classifiers.

Using the publicly available data sets associated with [1] and [2], we identified reliable classifiers that have feasible molecular implementations by limiting the scope of the classification problems and training support vector machines with high feature selection penalties. Regularization parameters were determined through iterative search, and a hard sparsity constraint of a maximum of 50 features was used. In particular, we identified a sparsely-featured classifier that differentiates between bacterial infections and viral infections from whole blood RNA samples, similar to that previously implemented as a molecular classifier for *in vitro* use [4]. We also created a sparsely-featured classifier that discriminates between cancer and healthy patients based on platelet RNA samples. Both classifiers were tested on validation sets that were disjoint from the training sets, and consisted of approximately $\frac{1}{6}$ the number of samples of the training sets, corresponding to validation sets of 35 and 56 samples respectively. The accuracy of the classifiers on the validation set will henceforth be referred to as model accuracy, and are shown in Table 1.

4 Simulating the Accuracy of a Molecular Classifier

The sequence dependent amplification bias in RASL is not well characterized, but based on the evaluation of probe bias in [12], we will roughly approximate it as a $\log_2$-normal distribution. The largest difference in amplification factor of the 25 probes tested in [12] is approximately 2^6, so we approximate the underlying normal distribution of probe bias to have $\mu = 0, \sigma = 3$.

We estimate the accuracy of a molecular implementation by generating random amplification biases according to the previously described distribution for each probe, then summing the biases of the probes for each target transcript in order to determine an effective weight for each feature. The resulting set of weights is then evaluated on the validation set to obtain an accuracy that can be compared to the model accuracy. Several such simulations are executed to determine an expected accuracy of a molecular implementation. The simulated accuracy of molecular classifiers implemented with the approach of [4] are shown in Table 1.

As mentioned before, the parameters of the distribution of probe bias are rough estimates. In Fig. 4, we see the effect of the standard deviation of the underlying normal distribution of probe bias on the simulated accuracy. This indicates that unless we have $\sigma < 1$, the accuracy of the molecular implementation of a classifier deviates highly from the model accuracy. Based on the data presented in [12], it seems unreasonable to expect that variation in amplification bias is so small. As such, a method of mitigating the effect of amplification bias is necessary.

Table 1. Properties of the sparsely-featured disease diagnostic classifiers. Basic molecular classifier accuracy is the estimated accuracy of the molecular classifiers implemented with amplification and without bias mitigation, averaged over 500 simulations.

Task	# of features	*In silico* model accuracy	Simulated basic molecular classifier accuracy
Viral vs. bacterial	5	0.89	0.59
Cancer diagnosis	24	0.96	0.60

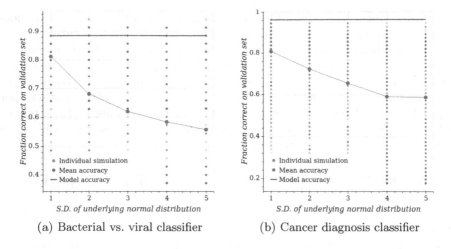

(a) Bacterial vs. viral classifier (b) Cancer diagnosis classifier

Fig. 4. Effect of variation in amplification bias of DNA probes on simulated accuracy of molecular classifiers implemented with no amplification bias mitigation. σ is the standard deviation of the normal distribution underlying the $\log_2$-normal distribution of probe bias, with a mean $\mu = 0$. For each σ, 500 independent simulations were done, and the shade of blue indicates the frequency of an outcome. (Color figure online)

5 Scaling up Weights to Reduce Bias Variation

The ideal method of mitigating amplification bias would involve no experimental procedures beyond those previously characterized in molecular gene expression classifiers. One such approach is to simply increase the number of probes used to implement a unit weight, such that a weight of magnitude k would be implemented by $c * k$ probes, where c is constant across all weights, as depicted in Fig. 5. This is equivalent to scaling up all weights by a constant factor. Intuitively, this will reduce the effect of probe bias because the average of c amplification factors should be less variable than the amplification factor of a single probe. The desired effect of this approach can be seen in the following mathematical characterization:

For a molecular classifier with a set of target transcripts $\mathcal{T}$, let

$$\mathcal{P}_t = \{p_1^t, p_2^t, \ldots p_{c*|t|}^t\}$$

be the set of amplification factors of the probes used to implement desired weight $|t|$ of a transcript $t \in \mathcal{T}$, where each p_i^t represents the amplification factor of the i-th probe targeting transcript t. We will define the effective weight of a target transcript t as

$$\mathcal{W}_t = \sum_{i=1}^{c*|t|} p_i^t$$

The effective weight is the weight implemented by the molecular classifier after amplification. Using the approximation that amplification factors are distributed

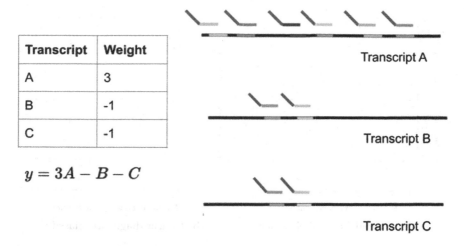

Transcript	Weight
A	3
B	-1
C	-1

$$y = 3A - B - C$$

Fig. 5. Molecular implementation of bias mitigation through scaling up weights, with scaling factor $c = 2$ for a simple toy classifier. Probes are not drawn to scale, and represent the RASL-seq probes shown in Fig. 3.

as $p_i^t \sim 2^{\mathcal{N}(\mu, \sigma)}$, we see that for large c,

$$\mathcal{W}_t \approx c * |t| * 2^\mu$$

In order to implement the desired classifier, the effective weights do not need to be the same as the desired weights – constantly scaled weights will yield identical classification results. As a result, if the effective weights have the same relative values as the desired weights, the implementation can be considered successful. We see that for any pair of transcripts $u, v \in \mathcal{T}$,

$$\frac{\mathcal{W}_u}{\mathcal{W}_v} \approx \frac{c * |u| * 2^n}{c * |v| * 2^n} = \frac{|u|}{|v|}$$

so the classifier will implement the desired weights, regardless of individual amplification biases.

However, c cannot be arbitrarily large, as it is limited by the number of unique binding sites per transcript and the cost of oligonucleotide synthesis. Instead, we must determine if an experimentally feasible c will yield a sufficient approximation for gene expression classification.

In Fig. 6, we see the effect of c on the simulated accuracy of a molecular classifier. The plots indicate that for even $c = 10$, corresponding to up to 50 distinct probes per transcript, there is still a large disparity between the predicted implementation accuracy and the model accuracy. Based on these results, it seems that knowledge of the amplification biases of the probes is necessary in order to construct a molecular classifier.

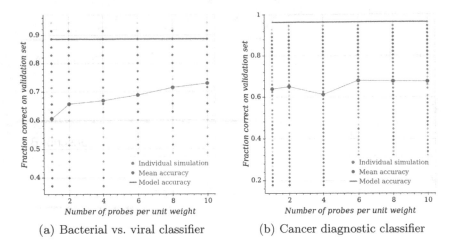

(a) Bacterial vs. viral classifier   (b) Cancer diagnostic classifier

Fig. 6. Simulated accuracy of molecular classifiers using c probes to implement a unit weight. The shade of blue indicates the frequency of an outcome. 500 independent simulations were done for each c. (Color figure online)

6   Classifier Calibration by Screening Probes

A simple extension of the approach to implementing molecular classifiers presented in [4] is to measure the amplification biases of a set of probes for each target transcript, then select a subset of these probes that will yield effective weights that best approximate the desired weights, as shown in Fig. 7. This approach can be mathematically characterized similarly to the approach described in Sect. 5:

For a molecular classifier with a set of target transcripts $\mathcal{T}$, let

$$\mathcal{P}_t = \{p_1^t, p_2^t, ... p_n^t\}$$

be the set of amplification factors of n probes targeting a transcript $t \in \mathcal{T}$ with desired weight $|t|$, where p_i^t represents the amplification factor of the i'th probe targeting transcript t. Let $\wp(\mathcal{P}_t)$ be the power set of $\mathcal{P}_t$, representing the set of all weights that could be implemented using the given set of probes. The effective weight of a subset of probes $\mathcal{S}_t \in \wp(\mathcal{P}_t)$ is

$$\mathcal{W}_t = \sum_{p_i^t \in \mathcal{S}_t} p_i^t$$

Our goal is to select subsets of probes $\mathcal{S}_t$ for all transcripts $t \in \mathcal{T}$ that yield

$$min\left(\sum_{u,v \in \mathcal{T}} \left(\frac{\mathcal{W}_u}{\mathcal{W}_v} - \frac{|u|}{|v|}\right)^2 \right)$$

This minimization problem requires us to search $|\wp(\mathcal{P}_t)|^{|\mathcal{T}|} = 2^{n*|\mathcal{T}|}$ possible solutions. For the bacterial vs. viral classifier, with an experimentally reasonable 10 probes per target transcript, there would be a massive 2^{50} possible sets

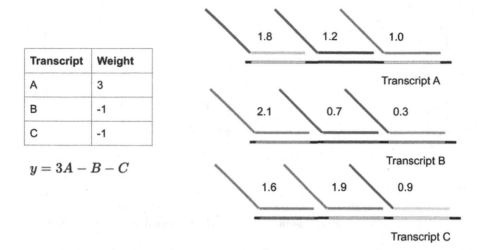

Transcript	Weight
A	3
B	-1
C	-1

$$y = 3A - B - C$$

Fig. 7. Molecular implementation of classifier built by selecting optimal subset of probes, with $n = 3$ probes per transcript, for a simple toy classifier, with probe biases (denoted by the number next to the probe) selected for demonstration purposes. The grey probes are omitted in the classifier implementation, and the colored probes are the subset that best approximates the desired weight. Probes are not drawn to scale, and represent the RASL-seq probes shown in Fig. 3. (Color figure online)

of weights, so it seems likely that the optimal selection of probes would yield desired classifier behavior. However, the large solution space also makes identifying the solution challenging. In order to estimate the accuracy of this approach analogously to Fig. 6, with 500 simulations of randomly biased probes, it is not feasible to search the entire solution space. Instead, for each transcript, we simply find the subset of probes whose sum is closest to the weight normalized to $n * \mu$, where μ is the mean of the $\log_2$-normal distribution. In other words, we select subsets of probes $\mathcal{S}_t$ for all transcripts $t \in \mathcal{T}$ that yield

$$min\left(\sum_{t \in \mathcal{T}} (\mathcal{W}_t - |t| * \mu)^2 \right)$$

This minimization problem requires us to search $|\wp(\mathcal{P}_t)| * |\mathcal{T}| = |\mathcal{T}| * 2^n$ possible solutions, which makes the search much more feasible for simulation purposes. The accuracy of the classifier built with this selection of probes gives us a reasonable lower bound on the accuracy of the optimal subset, which could feasibly be identified when implementing a disease diagnostic classifier. These estimated lower bounds of accuracy are shown for different n in Fig. 8. The lower bound of expected molecular classifier accuracy begins to approach the model accuracy at $n = 10$, which is experimentally feasible.

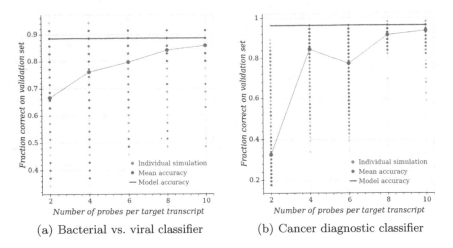

(a) Bacterial vs. viral classifier   (b) Cancer diagnostic classifier

Fig. 8. Simulated accuracies of molecular classifiers using subsets of n probes per transcript. The shade of blue indicates the frequency of an outcome. 500 independent simulations were done for each n. (Color figure online)

7   Encoding Weights in Probe Concentration

In the previously described methods of implementing molecular classifiers, in order to change the effective weights of the classifier, one must use a different set of probes to construct the classifier. It may be possible to instead encode classifier weights in probe concentration, similar to previous DNA-based neural network constructions [13]. Because disease diagnostic molecular classifiers must accept analog input, and we wish to minimize the number of strand displacement reactions in the system, the mechanism of weight multiplication should be different from previous DNA-based neural network constructions. An implementation of analog multiplication by DNA strand displacement has been proposed [16], but the design involves a complex reaction network that would likely be difficult to experimentally construct for the purposes of *in situ* molecular classification.

We have designed a method of implementing arbitrary classifier weights with a constant number of probes per transcript, by encoding classifier weights in the concentration of the probes. For each transcript, we select exactly one binding domain. For each binding domain, we design one positive probe and one negative probe. When constructing the classifier, we add all probes in excess, while controlling the ratios of each pair of positive and negative probes. The difference of the relative concentrations of the probes will encode the weight of the transcript. If a positive probe has the same concentration as its negative counterpart, then the weight of their target transcript should be 0. Similarly, a larger concentration of positive probe will encode a positive weight, and a larger concentration of negative probe will encode a negative weight. This weight implementation strategy is depicted in Fig. 9.

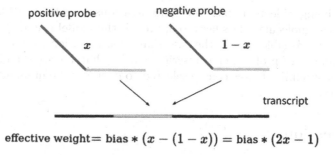

$$\textbf{effective weight} = \textbf{bias} * \big(x - (1 - x)\big) = \textbf{bias} * (2x - 1)$$

Fig. 9. Cartoon depiction of competitive hybridization based weight implementation strategy. The arrows indicate complementary domains. Probes are not drawn to scale, and represent the RASL-seq probes shown in Fig. 3. $x = \frac{|\text{positive probe}|}{|\text{positive probe}| + |\text{negative probe}|}$, the concentration of the positive probe normalized to the sum of the concentrations of the positive and negative probe.

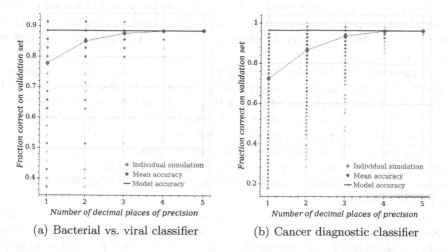

(a) Bacterial vs. viral classifier   (b) Cancer diagnostic classifier

Fig. 10. Simulated accuracies for molecular classifiers built with weights encoded in probe concentration. Decimals of precision indicates the control of the ratios. For example, 1 decimal of precision means that only ratios with increments of ten percent can be implemented.

Implementing weights in this fashion would allow for manipulation of classifier weights without changing the probes used. One could measure probe biases, then adjust the concentrations such that the effective weights are identical to the weights of the *in silico* model. Since the effective weight is simply a linear function of the relative concentrations, the ability to implement arbitrary weights is dependent on the ability to arbitrarily control the concentrations of probes. The simulated accuracy of molecular classifiers built with this weight implementation strategy after bias mitigation are shown in Fig. 10 for different degrees of concentration precision. We see that if ratios can be controlled to precision

10^{-3} (i.e. being able to implement concentrations in ratio 0.501:0.499), then we expect the molecular classifiers to approach the model accuracy. This seems experimentally feasible, but in the case that such precision cannot be attained, a larger number of probes per transcript can be characterized, and the probes with the most similar biases can be selected to decrease the minimum difference in concentration.

8   Discussion

Based on these analyses, both selecting subsets of probes and encoding classifier weights in probe concentration appear to be viable approaches to mitigating amplification bias in molecular classifiers. When mitigating bias by selecting subsets of probes, one can use almost exactly the experimental construction presented in [4]. The only additional steps involve characterizing probe bias. For this reason, it is nearly certain that such a strategy could be used to build molecular classifiers. However, it is possible that many probes will need to be screened in order to identify a subset of probes that implements the desired classifier.

While encoding the classifier weights in probe concentration requires a slightly different, currently experimentally unverified construction, it would likely use a much smaller set of probes in order to implement the desired classifier, and thus be more cost effective. Even with very limited implementation precision, one could imagine having to screen far fewer probes than in the subset selection method. Furthermore, it allows one to implement weights with more specificity than the subset selection method, so provided that there is sufficient implementation precision, it should allow for more accurate implementation of an *in silico* classifier.

Gene expression profiling is increasingly an important clinical metric for diagnosis a wide number of human diseases. The approach presented here would enable amplification and classification of RNA samples containing multiple biomarkers in a single reaction with a two-channel fluorescent readout. This could drastically reduce the complexity of gene expression classification and potentially enable a point-of-care solution to this type of diagnostics. Even though this approach would result in a more complex development stage, once a final construction is found, the implementation is drastically simpler than existing alternatives.

Acknowledgment. G. G. was supported by Caltech's Summer Undergraduate Research Fellowship program. R. L. and G. S. were supported by NSF grant CCF-1714497.

References

1. Tsalik, E.L., et al.: Host gene expression classifiers diagnose acute respiratory illness etiology. Sci. Transl. Med. **8**, 322ra11 (2016)
2. Best, M.G., et al.: RNA-Seq of tumor-educated platelets enables blood-based pancancer, multiclass, and molecular pathway cancer diagnostics. Cancer cell **28**, 666–676 (2015)
3. Benenson, Y., Gil, B., Ben-Dor, U., Adar, R., Shapiro, E.: An autonomous molecular computer for logical control of gene expression. Nature **429**, 423–429 (2004)
4. Lopez, R., Wang, R., Seelig, G.: A molecular multi-gene classifier for disease diagnostics. Nat. Chem. **10**, 746–754 (2018)
5. Zhang, D.Y.: Cooperative hybridization of oligonucleotides. J. Am. Chem. Soc. **133**, 1077–1086 (2010)
6. Turberfield, A.J., Mitchell, J., Yurke, B., Mills Jr., A.P., Blakey, M., Simmel, F.C.: DNA fuel for free-running nanomachines. Phys. Rev. Lett. **90**, 118102 (2003)
7. Dirks, R.M., Pierce, N.A.: Triggered amplification by hybridization chain reaction. Proc. Nat. Acad. Sci. USA **101**, 15275–15278 (2004)
8. Seelig, G., Yurke, B., Winfree, E.: Catalyzed relaxation of a metastable DNA fuel. J. Am. Chem. Soc. **128**, 12211–12220 (2006)
9. Zhang, D.Y., Turberfield, A.J., Yurke, B., Winfree, E.: Engineering entropy-driven reactions and networks catalyzed by DNA. Science **318**, 1121–1125 (2007)
10. Zhang, D.Y., Seelig, G.: DNA-based fixed gain amplifiers and linear classifier circuits. In: Sakakibara, Y., Mi, Y. (eds.) DNA 2010. LNCS, vol. 6518, pp. 176–186. Springer, Heidelberg (2011). https://doi.org/10.1007/978-3-642-18305-8_16
11. Chen, S.X., Seelig, G.: A DNA neural network constructed from molecular variable gain amplifiers. In: Brijder, R., Qian, L. (eds.) DNA 2017. LNCS, vol. 10467, pp. 110–121. Springer, Cham (2017). https://doi.org/10.1007/978-3-319-66799-7_8
12. Li, H., Qiu, J., Fu, X.D.: RASL-seq for massively parallel and quantitative analysis of gene expression. Curr. Protoc. Mol. Biol. **98**, 4–13 (2012)
13. Cherry, K.M., Qian, L.: Scaling up molecular pattern recognition with DNA-based winner-take-all neural networks. Nature **559**, 370 (2018)
14. Burchill, S.A., Perebolte, L., Johnston, C., Top, B., Selby, P.: Comparison of the RNA-amplification based methods RT-PCR and NASBA for the detection of circulating tumour cells. Br. J. Cancer **86**, 102 (2002)
15. Deng, R., Zhang, K., Sun, Y., Ren, X., Li, J.: Highly specific imaging of mRNA in single cells by target RNA-initiated rolling circle amplification. Chem. Sci. **8**, 3668–3675 (2017)
16. Song, T., Garg, S., Mokhtar, R., Bui, H., Reif, J.: Analog computation by DNA strand displacement circuits. ACS Synth. Biol. **5**, 898–912 (2016)
17. Stougaard, M., Juul, S., Andersen, F.F., Knudsen, B.R.: Strategies for highly sensitive biomarker detection by Rolling Circle Amplification of signals from nucleic acid composed sensors. Integr. Biol. **3**, 982–992 (2011)
18. Takahashi, H., Matsumoto, A., Sugiyama, S., Kobori, T.: Direct detection of green fluorescent protein messenger RNA expressed in Escherichia coli by rolling circle amplification. Anal. Biochem. **401**, 242–249 (2010)
19. Stahlberg, A., Krzyzanowski, P.M., Jackson, J.B., Egyud, M., Stein, L., Godfrey, T.E.: Simple, multiplexed, PCR-based barcoding of DNA enables sensitive mutation detection in liquid biopsies using sequencing. Nucleic Acids Res. **44**, e105–e105 (2016)

Reversible Computation Using Swap Reactions on a Surface

Tatiana Brailovskaya, Gokul Gowri$^{(\boxtimes)}$, Sean Yu, and Erik Winfree

California Institute of Technology, Pasadena, CA 91125, USA
{tbrailov,ggowri,ssyu,winfree}@caltech.edu

Abstract. Chemical reaction networks (CRNs) and DNA strand displacement systems have shown potential for implementing logically and physically reversible computation. It has been shown that CRNs on a surface allow highly scalable and parallelizable computation. In this paper, we demonstrate that simple rearrangement reactions on a surface, which we refer to as swaps, are capable of physically reversible Boolean computation. We present designs for elementary logic gates, a method for constructing arbitrary feedforward digital circuits, and a proof of their correctness.

1 Introduction

In traditional digital logic, information is lost, making computation logically irreversible. For example, an AND gate transforms two inputs into a single output, where the inputs cannot be deduced from the output. Landauer argued that irreversible logic, implemented by physically irreversible systems, dissipate at least a minimum energy, $kT \log 2$, with each binary computational step [1]. Charles Bennett countered that computation could be done in a logically reversible fashion, indicating that physically reversible systems could compute with arbitrarily little energy expenditure [2]. A surprising flip side to this discovery was that several simple models of constant-energy reversible systems, such as perfect billiard balls [3,4] and even a microscopic model of heat diffusion within molecular aggregation [5], were shown to be capable of carrying out arbitrary computations. The connections between computation and thermodynamics are now richly developed [6,7]. However, despite considerable effort [8], practical computing systems that perform reversible computing with asymptotically minimal energy expenditure have not been demonstrated.

In his seminal work, Bennett references biological nucleic acid systems as examples of logically and physically reversible computing [2,9]. Recently, building on techniques for compiling arbitrary formal chemical reaction networks (CRNs) into DNA strand displacement systems (DSDs) [10–12], the potential of synthetic nucleic acid systems to implement reversible computing has been explored theoretically and shown to be feasible for polynomial-space problems [13–15]. By further storing information in a DNA polymer based system,

---

T. Brailovskaya, G. Gowri and S. Yu—Equal contribution.

© Springer Nature Switzerland AG 2019
C. Thachuk and Y. Liu (Eds.): DNA 25, LNCS 11648, pp. 174–196, 2019.
https://doi.org/10.1007/978-3-030-26807-7_10

a scheme for physically and logically reversible Turing-universal computing has been proposed [16]. However, neither of these approaches allow for parallel computing, and they require many distinct species.

This motivated a framework for computing using chemical reaction networks on a surface (surface CRNs) [17]. In a surface CRN, a bimolecular reaction $A + B \to C + D$ can occur if a molecule of species A is adjacent to a molecule of species B on the surface. A C molecule will replace the A molecule, and a D molecule will replace the B molecule. Note that the molecules can be adjacent in any orientation, such that $A + B \to C + D \equiv B + A \to D + C$; both are distinct from $A + B \to D + C$. The surface CRN framework can be used to construct massively parallelizable space-bounded Turing machines and continuously active logic circuits of different sizes with a constant set of species [17].

In the proposed DNA implementation of surface CRNs [17], species are bound to a DNA origami surface, and free-floating fuel molecules are consumed (and waste molecules produced) to facilitate irreversible reactions between two neighboring species via DNA strand displacement [17]. While one could simulate a reversible surface CRN by utilizing pairs of irreversible reactions of the form $\{A + B \to C + D; C + D \to A + B\}$, it is plausible that a genuinely reversible implementation could be devised such that the waste of the forward reaction is the fuel of the reverse reaction, and vice versa (as is the case for some DSD implementations of well-mixed solution CRNs [16]). Implementing such a surface CRN in DNA would involve using one fuel molecule for the forward reaction, and a different fuel molecule for the reverse reaction. In a closed system, where there is no external power maintaining fuels and wastes at constant concentrations, an occurrence of the forward reaction would bias the system toward the reverse direction, as the amount of fuel for the forward direction decreases, while the amount of fuel for the reverse reaction increases. Computation using these pseudo-reversible reactions would be difficult to drive forward unless the system guarantees that each reaction is used equally in each direction, on average [13–15]. Such restrictions impose difficult design constraints.

Another approach to constructing reversible surface CRNs is to use only reactions of the form $A + B \to B + A$, which are implicitly reversible because surface CRNs do not consider absolute orientation. When the forward reaction's waste is the reverse reaction's fuel, there is therefore no net change in fuels or wastes – the implementation is effectively catalytic. These reactions are simply the rearrangement of two neighboring molecules, henceforth referred to as swaps, and abbreviated as $A \leftrightarrow B$.

Swap reactions were previously discussed as a way to simulate diffusion on a surface [17], but we have found that they can exhibit much more complex programmable behavior. One may initially think that systems built using only swap reactions cannot perform useful computation. No new species can be introduced after computation begins, so NOT gates and signal fanout may at first seem infeasible. Furthermore, since reactions are completely reversible, it may also seem that computation cannot be biased to proceed forward. Indeed, if swap reactions were implemented in a well-mixed solution rather than on a surface, they would be utterly useless.

However, because we are considering swap reactions on a surface, we can take advantage of the geometry of the surface and local nature of the swaps. We are able to obtain behavior that is far more controlled than random diffusion. In this paper, we will show that by carefully designing initial configurations and swap reactions, arbitrary digital logic circuits can be computed using reversible reactions on a surface.

2   Computing Paradigm

We seek to construct a set of species, along with a set of permissible swap rules and a starting arrangement of those species on the lattice, that is able to compute logical functions.

To represent bits, we will use two species, denoted 1 and 0. Each lattice point will be shown as a square. Within an arrangement of species on a surface, certain lattice points will be designated as input locations and certain lattice points will be designated as output locations. We will denote the input locations with arrows pointing towards them and outputs with arrows pointing away (e.g. see the arrows in the leftmost and rightmost locations in Fig. 1a). In our constructions, if the input lattice locations are replaced with any permutation of bits, then it is possible to reach an arrangement (through the permissible swaps) where the output locations are filled with bits. We can think of this abstractly as the arrangement computing some Boolean function $f : \{0,1\}^n \rightarrow \{0,1\}^m$, where n, m are the number of input and output lattice locations respectively, so long as a unique set of output values are reachable.

Consider the task of constructing a wire. Let's define a species w that is permitted to swap both with 0 and 1 (i.e. we permit the reactions $0 \leftrightarrow W$ and $1 \leftrightarrow W$). Imagine a line of W, as shown in Fig. 1.

If we let the leftmost lattice point be the input location and rightmost be the output location, then this arrangement computes the identity function. The input bit will randomly walk along the line of W, eventually occupying the output location. Any line of W is a wire, as it simply transmits a signal. In order to develop more complex systems, we will introduce more species and rules, and use larger layouts. With this paradigm in mind, we design composable logic gates, which allow us to construct arbitrary digital logic circuits.

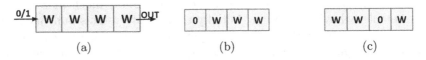

Fig. 1. Basic wire operation. **(a)** Wire starting configuration. **(b)** Initial loaded configuration. **(c)** Computation in progress.

3 Building Logic Circuits with Swaps

We have designed a constant set of 16 species and 23 swap rules, shown in Table 1, that are capable of universal Boolean logic, through the composition of NOT, AND, OR, fanout, and wirecross gate layouts on a surface.

Table 1. List of swap rules required to implement NOT, AND, OR, fanout and wirecross. No species beyond those that appear in these swap rules are necessary. Recall that the rule $A \leftrightarrow B$ is equivalent to the implicitly reversible surface CRN reaction $A + B \rightarrow B + A$

1. $0 \leftrightarrow W$	5. $0 \leftrightarrow I$	9. $I0 \leftrightarrow J$	13. $I0 \leftrightarrow X$	17. $I0 \leftrightarrow J0$
2. $1 \leftrightarrow W$	6. $1 \leftrightarrow I$	10. $I1 \leftrightarrow J$	14. $I1 \leftrightarrow X$	18. $I1 \leftrightarrow J1$
3. $0 \leftrightarrow W0$	7. $0 \leftrightarrow I0$	11. $W0 \leftrightarrow P$	15. $W0 \leftrightarrow X$	19. $W0 \leftrightarrow P0$
4. $1 \leftrightarrow W1$	8. $1 \leftrightarrow I1$	12. $W1 \leftrightarrow P$	16. $W1 \leftrightarrow X$	20. $W1 \leftrightarrow P1$
	21. $I \leftrightarrow J$	22. $K \leftrightarrow 1$	23. $K \leftrightarrow I$	

3.1 NOT, AND, OR

For these gates, we had several important goals in mind, the most important being that the gates had to be logically reversible. Fredkin and Toffoli demonstrate in [3] a set of universal logic gates that are reversible called the Fredkin and the Toffoli gates. Each of them is a three input - three output gate in which it is uniquely possible to determine the input from the output.

In the case of AND, OR logic, this is impossible (since the gate is not injective). However for our systems, it is sufficient to be able to uniquely determine the input to a gate from the output *plus* the final configuration of the gate. The idea is that even though different inputs give the same output, the way in which inputs rearrange in the gate will be unique.

NOT Logic. Initially, it may seem impossible to construct a NOT gate using only swap reactions, as the output of a NOT gate is not present in the input, and swap reactions cannot generate new species. However, we are able to implement a NOT gate by instead storing both a 1 and a 0 inside the gate, and releasing the appropriate bit depending on the input.

The starting configuration of a NOT gate is shown in Fig. 2a. In this configuration, no swaps are possible. Upon the arrival of a bit in the input location, it goes either right or left depending on its value, as 1 swaps with $W1$ and not $W0$, and 0 swaps with $W0$ and not $W1$. Upon this first transition, the bit may now swap with the I species, which may now travel down the two consecutive J species, and release the appropriate output bit. The output bit can then swap into the output location. This motif is symmetric for both inputs, as shown in Fig. 2b. An example trajectory for the 0 input case is shown in Fig. 2c.

Statespace. In order to give a more quantitative explanation of what is happening in this swap system, we have to introduce the notion of the state space as a graph. The nodes of said graph are unique configurations on the 2D surface. Two nodes are connected by an edge if a single swap can take one vertex to the other. From this point forward this graph will be referred to as the statespace graph. For instance, the state space graph for the NOT gate can be seen in Fig. 2d. It is simply a linear graph of 7 nodes, which corresponds to the sequence of states shown in Fig. 2c.

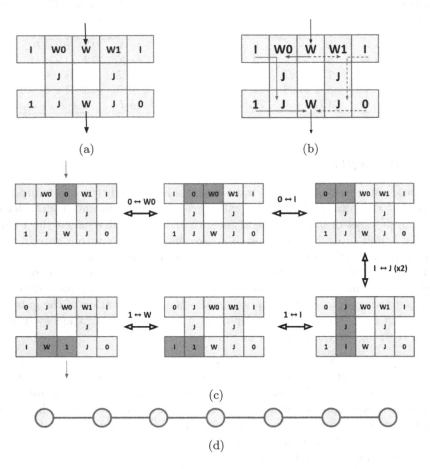

Fig. 2. NOT gate implementation via swap reactions. The unidirectional black arrows indicate the input and output locations. **(a)** Gate layout and swap rules. **(b)** Condensed representation of computation. Blue lines represent the initial movement of the input bit. The red represents the trajectory of the I species. The purple line is the movement of the output bit, which occurs after interacting with the I species. Dotted lines refer to when the input is 1, while solid lines are for input 0. **(c)** Computation trajectory on input 0. The bidirectional black arrows represent transition via swap rule. **(d)** Statespace of the NOT gate. (Color figure online)

All of our logic gates (AND, OR, NOT) have the property that there is only one state in which a correct output is in the output square. This means that in whatever circuit we embed our gates into, when the output leaves the gate, the remainder of the gate can only be in one possible configuration. In particular with the logic gates that we have constructed, after the output bit leaves, the remainder of the gate is stable (where stable means that no swaps can happen between the species in the remaining configuration). Furthermore, the gates are stable before any inputs arrive. This stability property is achieved without sacrificing reversibility by leveraging spatial separation.

AND Logic. Next, we will show how a small change in the NOT gate construction yields an AND gate. Consider a gate with the same layout as NOT except that the 1 and 0 species that are stored inside the gate are switched. If this gate receives an input 1, the output will be 1. Similarly, if the gate receives input 0, the output with be 0. Now, further modify this gate by removing the stored 1 and turning it into another input location, as shown in Fig. 3a. Observe that an AND gate will return 0 whenever there is a 0 in the input. Thus, in this construction, if the top input receives a 0, the gate will correctly output 0, regardless of what value arrives at the side input. It remains to verify that this gate works correctly when the top input receives a 1. Whenever the top input receives a 1, the side input will appear in the output location. Thus, if the side input receives 0, the output will be 0, as it should. Similarly, if the side input and top input both receive 1, the output will be 1, as desired.

OR Logic. Even though AND and NOT are sufficient for expressing any logical formula, it is convenient to also have a simple way of computing OR directly. Observe that in an OR gate, whenever there is a 1 in the input, the output should be 1. Thus, we can create OR analogously to AND, by fixing 1 as a stored input that always appears if the top input is 1. The layout is shown in Fig. 3b.

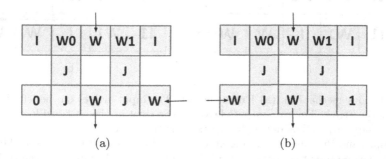

(a)   (b)

Fig. 3. Gate layout of AND and OR. For these two gates we have two inputs, denoted by the arrows pointing into the system. **(a)** AND gate. **(b)** OR gate.

Selector Motif. If no bits are stored in the previously described gates such that they have three input locations (top, left, and right), then the gate is equivalent to a data selector (or multiplexer), a device frequently used by electrical engineers that outputs one of two values determined by an additional input signal. We have shown the implementation of AND, OR, and NOT by fixing certain inputs, but we can also implement several other functions with this strategy, including strictly less than and strictly greater than. However, in our constructions, we use only the canonical AND, OR, and NOT gates.

The selector design was originally inspired by the Fredkin gate [3]. The Fredkin gate works by switching signals two and three in the output if signal one is 1, else the signals are output unchanged. A Fredkin gate can be programmed by fixing one or two of the three input signals and observing the output value of either signal two or three to compute AND, OR and NOT. Here, we effectively construct a Fredkin gate that only outputs the relevant signal that stores the value of AND, OR, or NOT.

3.2   Fanout, Wirecross

Although they are not essential for universality, wirecross and fanout are very helpful in creating compact circuits. It has been shown that it is possible to construct a wirecross from just fanout and elementary logic gates [18]. Wirecross seems challenging on a two dimensional surface, as wires cannot use the third

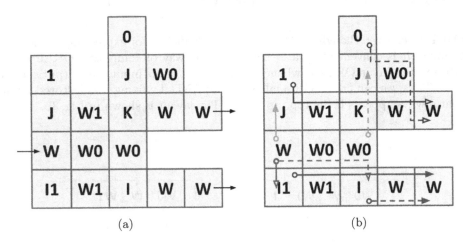

(a)   (b)

Fig. 4. Fanout mechanism. **(a)** Fanout layout. The input arrives on the left side indicated by the arrow. The two outputs leave on the right at the specified locations. **(b)** Fanout computation. The dotted lines represent the computation when a 0 bit serves as the input and the solid lines represent computation when a 1 bit serves as the input. The red lines represent the trajectory of the input. The yellow lines are the trajectory of the $I1$ and I species for the cases in which the input is 1 and 0 respectively. The green lines are the movement of the top output bit and the blue lines are the movement of the bottom output bit. (Color figure online)

dimension to avoid intersection. Fanout also seems challenging since we cannot create new species via swaps. We are able to implement wirecross by effectively constructing two unique wire types, and we are able to implement fanout by storing the additional bits inside the gate.

Fanout. In Fig. 4a we demonstrate our design for a fanout gate. The key idea is to store an additional copy of the bit inside the gate to simulate bit duplication. Overall, the gate consists of two interconnected columns, each of which is capable of copying a bit. One column will duplicate the input bit if it is 1 and the other will duplicate the input if it is 0. The bit is "duplicated" by freeing the extra stored bit of the same value.

If the input bit is a 1, then the first column duplicates it, and the two outputs travel on wires parallel to the $W0$ wire. We introduce a new species K that can both carry the 1 bit and the indicator I species. We also have to introduce a new species $I1$ to allow the 1 bit to duplicate. Refer to Table 1 for the associated swap rules.

Wirecross. Designing a wirecross that allows two bits to cross paths is challenging because if the signals are travelling along identical wire types, there is nothing stopping them from trading places and going down the wrong wires. In order to get around this issue, we translate the bits into intermediates that travel on distinct wire types and then release the appropriate bit once the crossing has taken place. We see this design in Fig. 5. At the top, we translate the input bit $a \in \{0,1\}$ to $Wa \in \{W0, W1\}$. This Wa then travels along a path specified by the P species. The other input b is translated to $Ib \in \{I0, I1\}$, which travels along a path specified by the J species. The center species is initially occupied

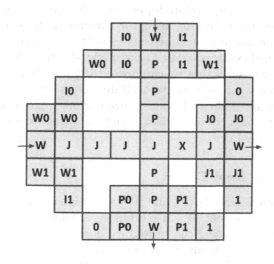

Fig. 5. Wirecross layout.

with a J, which means the input coming in from the left can pass, while the input from the top must wait. We introduce a final new species X which can swap with both Ib and Wa. First the Ib traverses the intersection and places the X species in the center. Then the displaced Wa can traverse the intersection point.

3.3   Compiling a Feedforward Circuit to a Surface Layout

With our gate constructions, we can compile arbitrary feedforward logic circuits (involving only AND, OR, NOT, fanout and wirecross) into surface layouts that evaluate the same functions via swap reactions. Here we describe one method of compilation of a circuit to a layout, following a standard crossbar array architecture [19]. This is not a spatially efficient way of constructing a surface layout; it is merely here to demonstrate that any feedforward circuit can be laid out on a surface.

We define a feedforward circuit as a graph by associating a circuit with a directed graph in the following way (for an explicit example of such a graph see Fig. 6a). Let the nodes denote inputs, outputs and gates (AND, OR, NOT). The in-edge at node b from node a indicates that output of node a is fed as input to the gate at node b. Similarly, an out-edge from node a to node b indicates that output of a is fed into the gate at node b. If such a graph is acyclic, then we say that the corresponding circuit is feedforward.

For simplicity, assume that all gate, fanout and wirecross layouts fit within a square of dimensions $s \times s$ lattice points, with sufficient space to route inputs on the top and left to outputs on the bottom and right. Our construction places the inputs regularly along the top of a rectangular region, and places the gates regularly along an offset diagonal, with inputs and gates each providing their output along vertical wires directly below them, extending to the bottom of the rectangle. To route the appropriate inputs to each logic gate, we use horizontal wires extending left from the gate, using the wirecross to pass through undesired wires and using the fanout to acquire the desired input signal while leaving that signal available for downstream gates that might also want it. The feedforward order of the circuit ensures that we can order the gates along the diagonal such that the inputs for a gate are always available to its left. As shown in Fig. 6b, this leads to a complete circuit layout of size not much larger than $(n + m)s \times 2ms$ lattice points for a circuit with n input and m gates, which is worst-case asymptotically optimal [20].

3.4   4-Bit Square Root Circuit

Following the examples in refs. [21] and [17], we designed a system that computes the floor of the square root of a 4-bit binary number. The layout, shown in Fig. 7, was not created using the procedure described in Sect. 3.3, but rather was designed by hand to be more compact. The correctness of the circuit was verified through exhaustive enumeration of the state space for every input, with long wires abbreviated in order to reduce simulation time. Since the system is

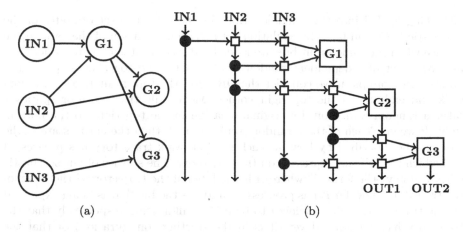

Fig. 6. A systematic feedforward circuit layout compilation. The circuit analyzed here computes $G2(G1(IN1, IN2), IN2)$ and $G3(G1(IN1, IN2), IN3)$. $G1$, $G2$, $G3$ are gates, $IN1$, $IN2$, $IN3$ are inputs, and $OUT1$, $OUT2$ are outputs. **(a)** Directed acyclic graph specification of the desired circuit. **(b)** Surface layout of the given circuit generated according to procedure described in Sect. 3.3. Filled circles indicate fanout. Empty squares indicate wirecross.

stochastic, parallel, and reversible, correctness was evaluated not with respect what the system *does* do, but rather with respect to what it *could* do, i.e. which states are reachable from a given input state. There was no combination of inputs for which the exhaustive enumeration found a state in which any incorrect output was produced, and for all input combinations the enumeration found a state in which all correct outputs were produced. Therefore, for all input combinations, the circuit will (with probability 1) eventually reach a state where all outputs are correct, and will never reach a state where any output is incorrect (though perhaps most of the time the outputs will be empty).

4   Proving Circuit Correctness

Leaving aside the question of kinetics for now, we will proceed to argue for the general correctness of feedforward circuit layouts, in the above sense. Our argument relies on establishing the composability of our gates in a surface CRN using strong and weak bisimulation similar to as in [22,23]. We start by showing that each gate is equivalent to a simple stochastic CRN. Then, we show that a composition of the stochastic CRNs for individual gates is equivalent to a composition of gates in a surface CRN. Finally, we will show that the composed stochastic CRN is logically equivalent to the intended function. In other words, when gates are composed on the surface CRN, the function they compute is exactly the logical composition of the gates.

When constructing an equivalent stochastic CRN for each gate in the surface CRN, we require that, upon starting with the set of species corresponding to bits being loaded into the input locations in the gate, the set of states of the stochastic CRN will be in bisimulation with the configuration of the gate on the surface when the same input bits are loaded. Observe that each gate's operation involves a set of independent random walks that come together at particular points. For example if we consider the OR gate, the top input takes a random walk and ends up in the top right corner. Meanwhile the indicator species I takes a random walk from its original location in the top right to the bottom right. If we give each of these random walks a label, then the entire state of the gate can be described by how far each random walk trajectory has progressed and which bit is being expressed on the trajectory. The species of our stochastic CRN will have the form A_i^b where A is the label of the trajectory, i the index of the trajectory (how far it has progressed) and b is the bit that is being expressed on the trajectory. We also allow i to be S or F indicating respectively that the trajectory has not started yet (it is in the starting configuration) or that the trajectory is finished.

The most basic reaction of the stochastic CRN is $A_i^b \leftrightarrow A_{i+1}^b$ indicating progress on trajectory A (provided A_{i+1}^b is an allowed species). When we have one trajectory lead unconditionally into another we see a reaction that looks like $A_n^b + B_S \leftrightarrow A_F^b + B_1^b$. When two trajectories come together in the surface CRN layout, this corresponds to reactions consuming the final numbered species of two trajectories and producing the first numbered species of a new trajectory.

In order for the stochastic CRN to be equivalent we must show that there is a bisimulation between the state space of the surface CRN (given a specific combination of inputs in the input locations) and the set of reachable states in the stochastic CRN (from the corresponding set of starting species). To demonstrate

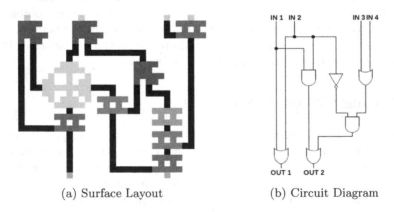

(a) Surface Layout   (b) Circuit Diagram

Fig. 7. 4-bit square root circuit. (**a**) Layout of the surface CRN 4-bit square root (done by hand in order to use less space than the general circuit layout scheme). Locations marked in purple represent the fanout mechanism. Red represents an OR gate, blue represents an AND, green a NOT, and grey a wirecross. Lattice points in light blue are input locations for the circuit and orange represents output locations. Black represents wires. (**b**) Abstract circuit diagram. (Color figure online)

this we will have a paint function f that maps states of the stochastic CRN to configurations (states) of the surface gate. This mapping will have the property that the available swaps from a state $f(s)$ on the surface correspond exactly to the set of reactions available to the stochastic CRN state s. By this we mean that applying the paint function f and then applying the swap will result in the same thing as applying the CRN reaction corresponding to the swap, and then applying the paint function. The paint function operates as follows, each species will specify how to paint a part of the surface. The parts of the surface that are not specified by any species will be the same as its initial configuration. We demonstrate a paint function for the OR gate in Fig. 8.

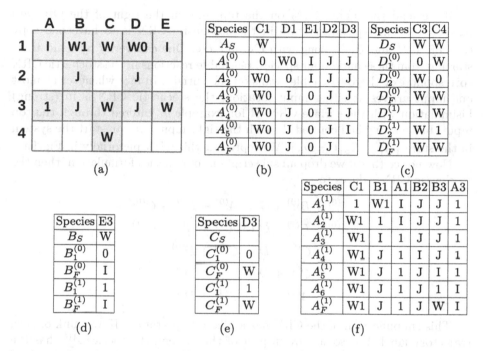

Fig. 8. Paint functions for the OR gate. All species in the stochastic CRN that correspond to the OR gate as well as their associated paint functions are shown. In each table, the left hand column is the list of species. Each row is a paint function for a particular species. Each column represents how a particular lattice point is painted. For example, the $A_F^{(0)}$ species would paint the lattice point C1 as $W0$, the lattice point D1 as J, and so on. **(a)** Initial configuration of the gate with each lattice point labeled with a row number and a column letter. **(b)** Paint function for $A^{(0)}$ including A_S. **(c)** Paint function for D. **(d)** Paint function for B. **(e)** Paint function for C. **(f)** Paint function for $A^{(1)}$ excluding A_S.

The reactions for the stochastic CRN that correspond to the OR gate are

$$A_5^{(0)} + B_1^{(0)} + C_S \leftrightarrow A_F^{(0)} + B_F^{(0)} + C_1^{(0)}$$
$$A_5^{(0)} + B_1^{(1)} + C_S \leftrightarrow A_F^{(0)} + B_F^{(1)} + C_1^{(1)}$$
$$C_1^{(0)} + D_S \leftrightarrow C_F^{(0)} + D_1^{(0)}$$
$$C_1^{(1)} + D_S \leftrightarrow C_F^{(1)} + D_1^{(1)}$$
$$A_6^{(1)} + D_S \leftrightarrow A_F^{(1)} + D_1^{(1)}$$
$$X_i^{(b)} \leftrightarrow X_{i+1}^{(b)} \quad \forall X \in \{A, D\}, \forall b \in \{0, 1\}, \forall i$$

$$(1)$$

If we load the inputs b_1, b_2 on the top and on the right of the OR gate respectively, the equivalent state in the stochastic CRN is to have one copy each of $A_1^{(b_1)}$ and $B_1^{(b_2)}$ along with C_S and D_S. One can verify by hand that, starting from these initial four species, reachable reactions in the stochastic CRN correspond exactly to reachable swaps on the surface. In fact whenever at most one input arrives at either input location, the stochastic CRN is in (strong) bisimulation with the surface CRN. If for example b_1 entered through the top input, but nothing had entered through the right input, we can start the system in the state $\{A_1^{(b_1)}, B_S, C_S, D_S\}$. We represent this CRN pictorially in Fig. 9a.

Now notice that if we drop all subscripts in our species formulation, then the stochastic CRN reduces to

$$A^{(0)} + B^{(0)} + C \leftrightarrow A^{(0)} + B^{(0)} + C^{(0)}$$
$$A^{(0)} + B^{(1)} + C \leftrightarrow A^{(0)} + B^{(1)} + C^{(1)}$$
$$C^{(0)} + D \leftrightarrow C^{(0)} + D^{(0)}$$
$$C^{(1)} + D \leftrightarrow C^{(1)} + D^{(1)}$$
$$A^{(1)} + D \leftrightarrow A^{(1)} + D^{(1)}$$

$$(2)$$

This simplification of the CRN has a nice interpretation. If we think of each trajectory label A as some wire in part of the system, the species $A^{(b)}$ has the interpretation of "wire A is hot and is carrying bit b". Now having an input of b_1, b_2 like described above is equivalent to starting with $\{A^{(b_1)}, B^{(b_2)}, C, D\}$. This reduced stochastic CRN is in weak bisimulation with the full stochastic CRN, for valid initial conditions. In Fig. 9b, we show a further simplified version of this CRN. The nodes in this graph represent distinct species (written out explicitly in Fig. 9a). Edges in this graph represent reactions with one node adjacent to that edge being a product and the other a reactant that additionally involves species C or D. Thus, Fig. 9b shows how different combinations of inputs $A^{(b_1)}$, $B^{(b_2)}$ are combined to obtain $D^{OR(b_1, b_2)}$. Thus this reduced stochastic CRN makes it easy to see that our gate works as expected, with A, B being the input trajectories and D being the output trajectory.

The equivalent paint functions for the stochastic CRNs for AND and NOT are straightforward extensions of the above. The equivalent paint function for the stochastic CRNs for them and for fanout and wirecross can be found in the Appendix. What happens now when we have a composition of gates in which the output lattice points of certain gates are linked with input lattice points of other gates via a (possibly bent) linear wire? We can create a "stitched" version of the stochastic CRN. We start with the stochastic CRN for each gate, and we relabel the species so that the same trajectories from two different OR gates, for example, are syntactically different. Next, we consider each wire in the composition. Suppose a given wire W links the output trajectory O of some gate to the input trajectory I of some other gate. Then we create a new trajectory for the movement of bits on a wire—let us call it W—and add the following reactions:

$$O_n^{(b)} + W_S \leftrightarrow O_F^{(b)} + W_1^{(b)}$$
$$W_l^{(b)} + I_S \leftrightarrow W_F^{(b)} + I_1^{(b)} \qquad (3)$$
$$W_i^{(b)} \leftrightarrow W_{i+1}^{(b)} \quad \forall b \in \{0,1\}, \forall i \text{ s.t. } 1 \leq i < l$$

where l is the length of the wire and $O_n^{(b)}$ is the last state of the output trajectory. In the reduced CRN this is effectively $O^{(b)} \leftrightarrow I^{(b)}$, which makes sense. Since we have already established every reaction in each gate's individual

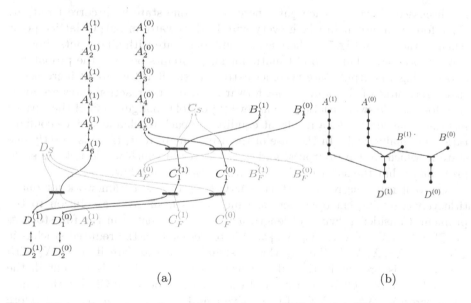

(a)   (b)

Fig. 9. Pictorial representation of the stochastic CRN for the OR gate.**(a)** The full stochastic CRN. Species and arrows shown in gray correspond to species that serve as indicators of whether a certain trajectory is in an initial or a final configuration. **(b)** A graph representation of the CRN without the grey indicator species. All species that only differ in their subscript are associated with each other. Note that the node above $D^{(1)}$ that appears to have two possible predecessors is actually deterministically reversible when the indicator species are accounted for.

stochastic CRN corresponds to some swap on the surface, its fairly clear that within this stitched stochastic CRN, all reactions still correspond to reachable swaps on the surface. What is left to show is that no reachable swaps cannot be expressed as some reaction in the stitched stochastic reaction. First, we claim that if the circuit is feedforward and each of its input locations is loaded with exactly one bit, then the circuit as a whole receives exactly 1 input per input wire. Consider what happens if this is not the case. Suppose a gate receives two inputs at an input lattice. Consider the first such occurrence. Let g be the first gate this happened to. g must have received two inputs from a previous gate, some g' that it was connected to. But g' could not have produced two outputs since we know g' received at most one input per lattice (by our selection of g). Therefore we have a contradiction, and so no such g can exist. If each gate will receive at most one input per input wire, the behavior of each gate is entirely captured by its individual stochastic CRN. Because the stochastic CRN for the circuit is a stitching of individual stochastic CRNs it must capture all possible swaps. Therefore the stitched stochastic CRN is in bisimulation with the surface CRN.

5   Kinetics and Entropics

As discussed above, for each gate there is only one state with correct output. Therefore, in a circuit in which every wire leads toward an output lattice point (such as the one in Fig. 7a), there is also only one state with all outputs simultaneously occupied. This is problematic for large circuits because the probability of observing an output state becomes extremely small as the circuit increases in size. The source of the issue is that forward and reverse reaction rates are identical for a single swap reaction. Thus, every possible configuration of the surface has the same energy. As a result, at equilibrium each reachable surface configuration is equally likely. In the case of the 4-bit square root, there is exactly one state in which we have outputs, and $> 10^6$ states in which we do not. As such, probability that the surface is in the output state is very low.

Conveniently, the reduced stochastic CRN provides a framework for quantitatively assessing entropic factors, which in turn suggests a solution to the problem. Consider a "wire" of length n that is represented in the full stochastic CRN by $\{X_S, X_1, \ldots, X_n, X_F\}$ and is represented in the reduced stochastic CRN by $\{X_S, X, X_F\}$. We say that a signal is on the wire if some X_i with $1 \leq i \leq n$ is present in the full stochastic CRN, and if X is present in the reduced stochastic CRN. A state α of the reduced stochastic CRN with signals on m wires $X^1, X^2, \ldots, X^m$ (of respective lengths $n_1, n_2, \ldots, n_m$) will therefore correspond to exactly $w_\alpha = \prod_{1 \leq k \leq m} n_k$ states of the full stochastic CRN. Consequently, we consider α to be a macrostate with Boltzmann weight w_α and thus equilibrium probability $p_\alpha = w_\alpha / Z$ where $Z = \sum_\alpha w_\alpha$. Equivalently, we could say that macrostate α has energy $E_\alpha = -kT \sum_{1 \leq k \leq m} \log n_k$.

As an initial application of this quantitative framework, we can determine the effect of the "unnecessary" fanout gates in the construction of Fig. 6b. These gates were introduced in the construction as a convenience so that input signals and computed signals can be propagated down vertical wires, just in case they will be needed later on – and we did not bother to complicate the construction by specifying how to remove the unnecessary wires and fanout gates. Yet they have an interesting side effect: system states with signals on all the output locations will also have signals trapped on the unnecessary wires. Thus, rather than there being exactly one such output state, there will be many—for example, the product of the lengths of the unnecessary wires, in a simple case. These wires therefore, by their presence, bias the computation forward. By introducing additional unnecessary fanout gates throughout the circuit, we could further bias each step of the computation to ensure that there are no deep valleys in the macrostate energy landscape that could kinetically trap the system. Noting that a fanout gate can reverse itself only if *both* its output signals come back, we can see that it provides a probabilistic kinetic ratchet as well.

As a more compact alternative to this use of fanout gates for their entropic side-effects, we introduce a compact entropic driver mechanism (Fig. 10) that biases computation forward. The mechanism results in an inflation of the number of states after a bit has traversed the mechanism.

Consider loading the gadget in Fig. 10 with a 0 bit species on the left. There is only one state in which there is a 0 bit species on the leftmost lattice point. However, if that bit travels to the right side, then we observe that the number of states in which the 0 bit species is on the rightmost point is 4. This is because after the bit travels to the right side, each of the I and J species and can swap; the entropy gadget has been "activated". Thus each of the pairs can either swap or not, which means we have $2 \times 2 = 4$ possible configurations.

If we have an entropic gadget placed after a gate G_1 connected via a wire to gate G_2, the output bit is less likely to return into G_1 and more likely to enter G_2, than if there was no entropic gadget present. This happens for two reasons. First, the J species could be blocking the wire for the bit to move backwards, favoring progression of computation forward. Second, the presence of the entropic gadget quadruples the number of states in which the bit has entered G_2, thus increasing the probability that the bit is in G_2. Thus, the entropic gadget can be used to increase the likelihood of observing the output state of a circuit.

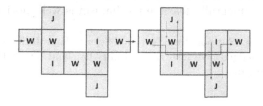

Fig. 10. The entropic driver gadget. The blue line indicates the trajectory of the primary bit traveling down the wire. The red and green lines represent the movement of the I and J species after the primary bit has passed. (Color figure online)

By adjusting the size or the number of entropic driver mechanisms, we can control the increase in number of states, and bias computation forward to an arbitrary degree. For instance, if we change the number of Js from 1 to n by adding more Js on top and bottom, on the left and the right of the entropic gadget, respectively, the number of states of the entropic gadget will be $(n + 1)^2$. With larger n, we further increase the likelihood that computation moves forward. Alternatively, we can put many entropic gadgets next to each other. A wire consisting of n gadgets in series will have 4^n states at the end of the wire. As computation proceeds, more entropy gadgets become activated and the number of possible configurations is exponential in the number of gadgets activated. Thus, by including sufficiently many entropy gadgets or increasing the number of J species, we can drive computation forward to an arbitrary extent.

By judiciously introducing entropy driver gadgets, it should be possible to modify any circuit such that the energies of the macrostates are decreasing roughly linearly as a function of number of gates completed, and thus computation is thermodynamically driven forward. Note that the entropy thus produced by circuit operation entails a corresponding energetic cost to performing the computation.

6   Future Directions

In this work, we constructed arbitrary feedforward circuits using reversible swap reactions on a surface. We also devised a entropic driver mechanism to tunably drive computation forward. However, many fascinating questions regarding swap reaction systems remain unanswered. For instance, the circuits we construct above are not reusable. Might it be possible to devise renewable swap systems that, for example, implement reversible sequential logic or reversible Turing machines using swap reactions?

When we first thought about swap reactions, a primary motivation was that due to the simplicity of such a reaction, there may exist a simple molecular implementation. The DNA strand displacement mechanism that was originally proposed for implementing arbitrary bimolecular reactions on a surface [17] is complex and potentially not suitable for experimental implementation. Now that the computational potential of swap rules has been established, the search for a simpler, more experimentally feasible mechanism is well motivated.

Acknowledgements. Support from National Science Foundation grant CCF-1317694 is gratefully acknowledged. We also thank Lulu Qian and Chris Thachuk for helpful discussion and comments.

Appendix

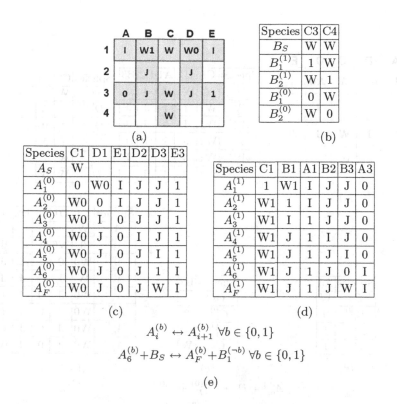

(a)

Species	C3	C4
B_S	W	W
$B_1^{(1)}$	1	W
$B_2^{(1)}$	W	1
$B_1^{(0)}$	0	W
$B_2^{(0)}$	W	0

(b)

Species	C1	D1	E1	D2	D3	E3
A_S	W					
$A_1^{(0)}$	0	W0	I	J	J	1
$A_2^{(0)}$	W0	0	I	J	J	1
$A_3^{(0)}$	W0	I	0	J	J	1
$A_4^{(0)}$	W0	J	0	I	J	1
$A_5^{(0)}$	W0	J	0	J	I	1
$A_6^{(0)}$	W0	J	0	J	1	I
$A_F^{(0)}$	W0	J	0	J	W	I

(c)

Species	C1	B1	A1	B2	B3	A3
$A_1^{(1)}$	1	W1	I	J	J	0
$A_2^{(1)}$	W1	1	I	J	J	0
$A_3^{(1)}$	W1	I	1	J	J	0
$A_4^{(1)}$	W1	J	1	I	J	0
$A_5^{(1)}$	W1	J	1	J	I	0
$A_6^{(1)}$	W1	J	1	J	0	I
$A_F^{(1)}$	W1	J	1	J	W	I

(d)

$$A_i^{(b)} \leftrightarrow A_{i+1}^{(b)} \ \forall b \in \{0,1\}$$

$$A_6^{(b)} + B_S \leftrightarrow A_F^{(b)} + B_1^{(\neg b)} \ \forall b \in \{0,1\}$$

(e)

Fig. 11. Paint functions for the NOT gate. All species in the stochastic CRN that corresponds to the NOT gate as well as their associated paint functions are shown. In each table, the left hand column is the list of species. Each row is a paint function for a particular species. Each column represents how a particular lattice is painted. **(a)** Initial configuration of the NOT gate with each lattice point labeled with a row number and a column letter. **(b)** Paint function for B. **(c)** Paint function for $A^{(0)}$ including A_S. **(d)** Paint function for $A^{(1)}$ excluding A_S. **(e)** Stochastic CRN equivalent to the NOT gate.

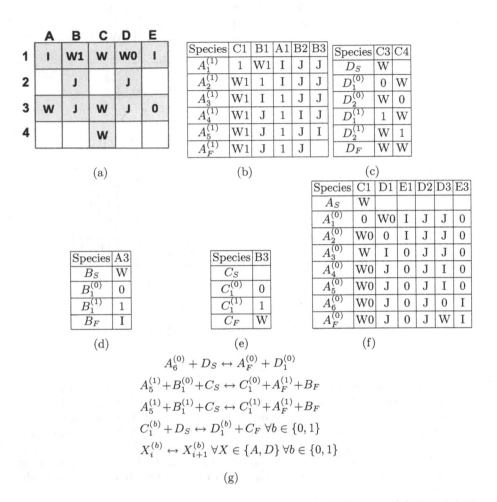

Fig. 12. Paint functions and the stochastic CRN for the AND gate. (a) Initial AND gate surface layout with each lattice labeled with a row number and a column letter. (b) Paint function for $A^{(1)}$. (c) Paint function for D. (d) Paint function for B. (e) Paint function for C. (f) Paint function for $A^{(0)}$. (g) Stochastic CRN equivalent to AND gate.

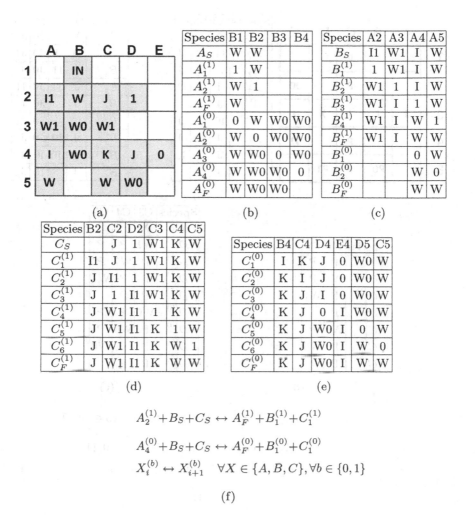

(a)

	A	B	C	D	E
1		IN			
2	I1	W	J	1	
3	W1	W0	W1		
4	I	W0	K	J	0
5	W		W	W0	

(b)

Species	B1	B2	B3	B4
A_S	W	W		
$A_1^{(1)}$	1	W		
$A_2^{(1)}$	W	1		
$A_F^{(1)}$	W			
$A_1^{(0)}$	0	W	W0	W0
$A_2^{(0)}$	W	0	W0	W0
$A_3^{(0)}$	W	W0	0	W0
$A_4^{(0)}$	W	W0	W0	0
$A_F^{(0)}$	W	W0	W0	

(c)

Species	A2	A3	A4	A5
B_S	I1	W1	I	W
$B_1^{(1)}$	1	W1	I	W
$B_2^{(1)}$	W1	1	I	W
$B_3^{(1)}$	W1	I	1	W
$B_4^{(1)}$	W1	I	W	1
$B_F^{(1)}$	W1	I	W	W
$B_1^{(0)}$			0	W
$B_2^{(0)}$			W	0
$B_F^{(0)}$			W	W

(d)

Species	B2	C2	D2	C3	C4	C5
C_S		J	1	W1	K	W
$C_1^{(1)}$	I1	J	1	W1	K	W
$C_2^{(1)}$	J	I1	1	W1	K	W
$C_3^{(1)}$	J	1	I1	W1	K	W
$C_4^{(1)}$	J	W1	I1	1	K	W
$C_5^{(1)}$	J	W1	I1	K	1	W
$C_6^{(1)}$	J	W1	I1	K	W	1
$C_F^{(1)}$	J	W1	I1	K	W	W

(e)

Species	B4	C4	D4	E4	D5	C5
$C_1^{(0)}$	I	K	J	0	W0	W
$C_2^{(0)}$	K	I	J	0	W0	W
$C_3^{(0)}$	K	J	I	0	W0	W
$C_4^{(0)}$	K	J	0	I	W0	W
$C_5^{(0)}$	K	J	W0	I	0	W
$C_6^{(0)}$	K	J	W0	I	W	0
$C_F^{(0)}$	K	J	W0	I	W	W

$$A_2^{(1)} + B_S + C_S \leftrightarrow A_F^{(1)} + B_1^{(1)} + C_1^{(1)}$$

$$A_4^{(0)} + B_S + C_S \leftrightarrow A_F^{(0)} + B_1^{(0)} + C_1^{(0)}$$

$$X_i^{(b)} \leftrightarrow X_{i+1}^{(b)} \quad \forall X \in \{A, B, C\}, \forall b \in \{0, 1\}$$

(f)

Fig. 13. Paint functions for the fanout gate. **(a)** Initial configuration of the fanout gate with each lattice labeled with a row number and a column letter. **(b)** Paint function for A. **(c)** Paint function for B. **(d)** Paint function for $C^{(1)}$ including C_S. **(e)** Paint function for $C^{(0)}$ excluding C_S. **(f)** Stochastic CRN equivalent to fanout.

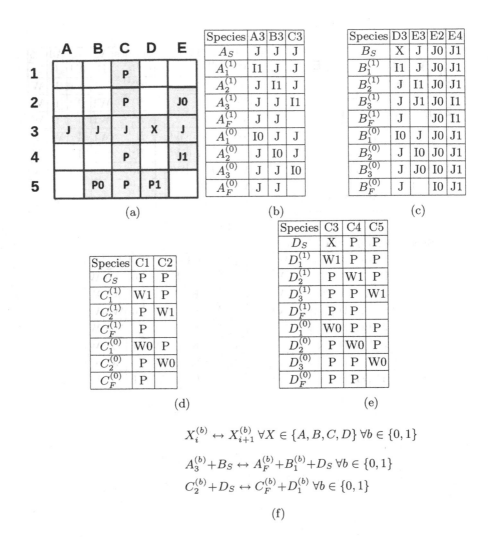

Fig. 14. Paint function for partial wirecross. Truncated portions are equivalent to trajectories found in AND/OR/NOT gates. (a) Initial configuration of the central portion of wirecross with each lattice labeled with a row number and a column letter. (b) Paint function for A. (c) Paint function for B. (d) Paint function for C. (e) Paint function for D. (f) Stochastic CRN equivalent to central portion of wirecross.

References

1. Landauer, R.: Irreversibility and heat generation in the computing process. IBM J. Res. Dev. **5**, 183–191 (1961)
2. Bennett, C.H.: Logical reversibility of computation. IBM J. Res. Dev. **17**, 525–532 (1973)
3. Fredkin, E., Toffoli, T.: Conservative logic. Int. J. Theor. Phys. **21**, 219–253 (1982)
4. Margolus, N.: Physics-like models of computation. Phys. D Nonlinear Phenom. **10**, 81–95 (1984)
5. D'Souza, R.M., Homsy, G.E., Margolus, N.H.: Simulating digital logic with the reversible aggregation model of crystal growth. In: Griffeath, D., Moore, C. (eds.) New Constructions in Cellular Automata, pp. 211–230. Oxford University Press, Oxford (2003)
6. Ouldridge, T.E.: The importance of thermodynamics for molecular systems, and the importance of molecular systems for thermodynamics. Nat. Comput. **17**, 3–29 (2018)
7. Wolpert, D.: The stochastic thermodynamics of computation. J. Phys. Math. Theor. **52**, 193001 (2019)
8. Perumalla, K.S.: Introduction to Reversible Computing. Chapman and Hall/CRC, Boca Raton (2013)
9. Bennett, C.H.: The thermodynamics of computation-a review. Int. J. Theor. Phys. **21**, 905–940 (1982)
10. Soloveichik, D., Seelig, G., Winfree, E.: DNA as a universal substrate for chemical kinetics. Proc. Natl. Acad. Sci. **107**, 5393–5398 (2010)
11. Chen, Y.-J., et al.: Programmable chemical controllers made from DNA. Nat. Nanotechnol. **8**, 755–762 (2013)
12. Srinivas, N., Parkin, J., Seelig, G., Winfree, E., Soloveichik, D.: Enzyme-free nucleic acid dynamical systems. Science **358**, eaal2052 (2017)
13. Thachuk, C., Condon, A.: Space and energy efficient computation with DNA strand displacement systems. In: Stefanovic, D., Turberfield, A. (eds.) DNA 2012. LNCS, vol. 7433, pp. 135–149. Springer, Heidelberg (2012). https://doi.org/10.1007/978-3-642-32208-2_11
14. Codon, A., Hu, A.J., Manuch, J., Thachuk, C.: Less haste, less waste: on recycling and its limits in strand displacement systems. Interface Focus **2**, 512–521 (2012)
15. Condon, A., Thachuk, C.: Towards space- and energy-efficient computations. In: Kempes, C., Grochow, J., Stadler, P., Wolpert, D. (eds.) The Energetics of Computing in Life and Machines, Chap. 9, pp. 209–232. The Sante Fe Institute Press, Sante Fe (2019)
16. Qian, L., Soloveichik, D., Winfree, E.: Efficient turing-universal computation with DNA polymers. In: Sakakibara, Y., Mi, Y. (eds.) DNA 2010. LNCS, vol. 6518, pp. 123–140. Springer, Heidelberg (2011). https://doi.org/10.1007/978-3-642-18305-8_12
17. Qian, L., Winfree, E.: Parallel and scalable computation and spatial dynamics with DNA-based chemical reaction networks on a surface. In: Murata, S., Kobayashi, S. (eds.) DNA 2014. LNCS, vol. 8727, pp. 114–131. Springer, Cham (2014). https://doi.org/10.1007/978-3-319-11295-4_8
18. Goldschlager, L.M.: The monotone and planar circuit value problems are log space complete for P. ACM SIGACT News **9**, 25–29 (1977)
19. Masson, G.M., Gingher, G.C., Nakamura, S.: A sampler of circuit switching networks. Computer **12**, 32–48 (1979)

20. Savage, J.E.: Models of Computation. Addison-Wesley, Reading (1998). Section 12.6
21. Qian, L., Winfree, E.: Scaling up digital circuit computation with DNA strand displacement cascades. Science **332**, 1196–1201 (2011)
22. Milner, R.: Communication and Concurrency. Prentice Hall, Upper Saddle River (1989)
23. Johnson, R.F., Dong, Q., Winfree, E.: Verifying chemical reaction network implementations: a bisimulation approach. Theor. Comput. Sci. **765**, 3–46 (2019)

Error-Free Stable Computation with Polymer-Supplemented Chemical Reaction Networks

Allison Tai[✉] and Anne Condon

University of British Columbia, Vancouver, BC V6T 1Z4, Canada
{tyeuyang,condon}cs.ubc.ca

Abstract. When disallowing error, traditional chemical reaction networks (CRNs) are very limited in computational power: Angluin et al. and Chen et al. showed that only semilinear predicates and functions are stably computable by CRNs. Qian et al. and others have shown that polymer-supplemented CRNs (psCRNs) are capable of Turing-universal computation. However, their model requires that inputs are pre-loaded on the polymers, in contrast with the traditional convention that inputs are represented by counts of molecules in solution. Here, we show that psCRNs can stably simulate Turing-universal computations even with solution-based inputs. However, such simulations use a unique "leader" polymer per input type and thus involve many slow bottleneck reactions. We further refine the polymer-supplemented CRN model to allow for anonymous polymers, that is, multiple functionally-identical copies of a polymer, and provide an illustrative example of how bottleneck reactions can be avoided in this new model.

Keywords: Stable computation · Chemical reaction networks · DNA polymers

1 Introduction

The logical, cause-and-effect nature of chemical reactions has long been recognized for its potential to carry information and make decisions. Indeed, biological systems exploit interacting digital molecules to perform many important processes. Examples include inheritance with DNA replication, passing information from the nucleus to the cytoplasm using messenger RNA, or activating different cellular states through signal transduction cascades. Chemical reaction networks (CRNs) exploit these capabilities to perform molecular computations, using a finite set of molecular species (including designated input and output species) and reactions. Reactions among molecules in a well-mixed solution correspond to computation steps. CRN models may use mass-action kinetics, where the dynamics of the reactions are governed by ordinary differential equations, or stochastic kinetics, where the choice of reaction and length of time between reactions depends on counts of molecular species. We focus on stochastic CRNs in this work.

© Springer Nature Switzerland AG 2019
C. Thachuk and Y. Liu (Eds.): DNA 25, LNCS 11648, pp. 197–218, 2019.
https://doi.org/10.1007/978-3-030-26807-7_11

Stochastic CRNs using unbounded molecular counts are Turing-universal [1], but have a non-zero chance of failure. One challenge is that CRNs are unable to detect the absence of a molecule, and therefore when all molecules of a particular species have been processed. For example, when trying to simulate a register machine, if their count of a species corresponds to a register value, then a test-for-zero instruction needs to detect when all molecules have been depleted, i.e., their count is zero. Indeed, while the error of a CRN simulation of Turing-universal computation can be made arbitrarily small, it can never reach zero [2].

Error-free CRNs include those that exhibit *stable computation*: the output can change as long as it eventually converges to the correct answer; and *committing computation*: the presence of a designated "commit" species indicates that the output is correct and does not change subsequently. The class of predicates stably computable by CRNs is limited to semilinear predicates [3], and functions computable by committing CRNs are just the constant functions [2].

Cummings et al. [4] introduced the notion of *limit-stable computation*, which relaxes the stability requirement. In a computation of a limit-stable CRN, the output may change repeatedly, but the probability of changing the output from its correct value goes to zero in the limit. Cummings et al. show that any halting register machine can be simulated by a limit-stable CRN. Their construction involves repeated simulations of a register machine, resetting the register values each time, along with slowing down any error-prone reactions each time they occur. They show that the computational power then becomes equivalent to that of a Turing machine with the ability to change its output a finite number of times, capable of deciding predicates in the class Δ_2^0 of limit-computable function.

From these insights, we can see that CRN computations that produce the correct answer with probability 1 are still severely limited. We ask, "Is there any way to extend CRNs to work around the lack of ability to detect absence?" Qian et al. [5] gave a promising answer to this question by introducing a CRN model that is supplemented by polymers that behave as stacks, onto which monomers can be pushed and popped. Most importantly, this extended model allows for the stack base unit $\perp$ to be used as a reactant in implementing a "stack empty" operation. Indeed, Qian et al. use this operation in an error-free simulation of a stack machine. The resulting protocol, however, requires that the entire input is pre-loaded onto one of the stacks, a large change from traditional CRNs which assumes the inputs are well-mixed in a solution.

Motivated by the work of Qian et al., we wish to go one step further: Is Turing-universal stable computation by polymer-supplemented CRNs (psCRNs) possible when the input is represented by counts of monomers in a well-mixed solution? Intuitively, if input monomers can be loaded on to a polymer, absence of that species from the system could be detected by emptiness of the polymer, and thus circumvent a significant barrier to error-free, Turing-universal computation. The obvious obstacle is that it seems impossible to guarantee that all inputs are loaded on the polymer before computation can begin, if we can't reliably check for absence of inputs in the environment. At first glance, the logic appears circular, but we show that indeed stable Turing-universal computation is possible, and also present ideas for speeding up such computations.

In the rest of this section we describe our four main contributions and review related work. Section 2 introduces our polymer CRN models, Sects. 3, 4, and 5 describe our results, and Sect. 6 concludes with a summary and directions for future work.

1.1   Contributions and Highlights

Stable Register Machine Simulation Using CRNs with Leader Polymers. We design a polymer-supplemented CRN (psCRN) that simulates a register machine, assuming that all inputs are pre-loaded on polymers, with one polymer per species. We then augment the simulator CRN with CRNs that detect when an input has been loaded, and that restarts the simulator in this case. This scheme is similar to the error correction scheme of Cummings et al., but leverages polymers to ensure stable computation. Our polymer simulation of register machines, and thus Turing-universal computation, has a unique polymer per input species, as well as a "program counter" to ensure that execution of reactions follows the proper order. In the parlance of traditional CRNs, these molecules serve as "leaders". As a consequence, the simulation has a high number of so-called bottleneck reactions, which involve two leader reactants. Bottleneck reactions are undesirable because they are slow.

Anonymous Polymers Can Help Avoid Bottleneck Reactions. To avoid bottleneck reactions, we propose a CRN polymer model with no limit on the number of polymers of a given species, other than the limit posed by the volume of the system. In addition to type-specific increment, decrement and test-if-empty operations (which are applied to one polymer of a given species), polymer stubs can be created or destroyed. We call such polymers "anonymous". We illustrate the potential of psCRNs with anonymous polymers to reduce bottleneck reactions, by describing psCRN to compute $f(n) = n2^{\lfloor \lg n \rfloor}$ (which is the same as n^2 when n is a power of 2).

Abstractions for Expressing CRN Multi-threading and Synchronization. Our CRN for $f(n) = n2^{\lfloor \lg n \rfloor}$ uses threading to ensure that polymer reactions can happen in parallel, and uses a "leader" polymer for periodic synchronization. To describe our psCRN, we develop threading abstractions for psCRNs with anonymous polymers.

Time Complexity and a Simulator for CRNs with Anonymous Polymers. To test the correctness of our psCRNs and evaluate their running times, we developed a custom CRN simulator designed to support anonymous polymers and their associated reactions. Underlying our simulator is a stochastic model of psCRN kinetics that is a natural extension of traditional stochastic CRNs and population protocols. We also use this model to analyze the expected time complexities of our psCRNs examples in this paper, showing how speedups are possible with anonymous polymers.

1.2   Related Work

Soloveichik et al. [6] demonstrated how Turing-universal computation is possible with traditional stochastic CRNs, achieving arbitrarily small (but non-zero) error probability. For the CRN model without polymers Cummings et al. [4] showed how to reset computations so as to correct error and achieve limit-stable computation (which is weaker than stable computation).

In order to understand the inherent energetic cost of computation, Bennett [7,8] envisioned a polymer-based chemical computer, capable of simulating Turing machines in a logically reversible manner. Qian et al. [5] introduced a stack-supplemented CRN model in which inputs are pre-loaded on stacks, and showed how the model can stably simulate stack machines. Johnson et al. [9] introduce a quite general linear polymer reaction network (PRN) model for use with simulation and verification, as opposed to computation. Cardelli et al. [10] also demonstrated Turing-universal computation using polymers, using process algebra systems, but these systems are not stochastic. Jiang et al. [11] also worked on simulating computations with mass-action chemical reactions, using a chemical clock to synchronize reactions and minimize errors.

Lakin et al. [12] described polymerizing DNA strand displacement systems, and showed how to model and verify stack machines at the DSD level. They also simulated their stochastic systems using a "just-in-time" extension of Gillespie's algorithm. Their model has a single complex to represent a stack. Recognizing limitations of this, they noted that "it would be desirable to invent an alternative stack machine design in which there are many copies of each stack complex...", which is what we do in this paper. They propose that updates to stacks could perhaps be synchronized using a clock signal such as that proposed by Jiang et al. [11]. In contrast, our synchronization mechanism is based on detection of empty polymers.

The population protocol (PP) model introduced by Angluin et al. [13], which is closely related to the CRN model, focuses on pairwise-interacting agents that can change state. In Angluin et al.'s model, agents in a PP are finite-state. An input to a computation is encoded in the agents' initial states; the number of agents equals the input size. Any traditional CRN can be transformed into a PP and vice versa. Chatzigiannakis et al. [14] expand the n agents to be Turing machines, then examine what set of predicates such protocols can stably compute using $O(\log n)$ memory. Although the memory capacity of our polymers can surpass $O(\log n)$, polymer storage access is constrained to be that of a counter or stack, unlike the model of Chatzigiannakis et al.

2   Polymer-Supplemented Chemical Reaction Networks

A polymer-supplemented stochastic chemical reaction network (psCRN) models the evolution of interacting molecules in a well-mixed volume, when monomers can form polymers. We aim for simplicity in our definitions here, providing just enough capability to communicate the key ideas of this paper. Many aspects of our definitions can be generalized, for example by allowing multiple monomer species in a polymer, or double-end polymer extensibility, as is done in the work of Johnson et al. [9], Lakin et al. [12], Qian et al. [5], and others.

Reactions. A traditional CRN describes reactions involving molecules whose species are given by a finite set Σ. A *reaction*

$$r + r' \longrightarrow p + p' \tag{1}$$

describes what happens when two so-called reactant molecules of species $r, r' \in \Sigma$ collide: they produce molecules of species $p \in \Sigma$ and $p' \in \Sigma$. We assume that all reactions have exactly two reactants and two products, that the multi-sets $\{r, r'\}$ and $\{p, p'\}$ are not equal, that for any r and r', there is at most one reaction with reactants of species r and r'. For now we do not ascribe a rate constant to a reaction; we will do that in Sect. 5.

Polymer-supplemented chemical reaction networks (psCRNs) also have reactions pertaining to polymers. A designated subset $\Sigma^{(m)}$ of Σ is a set of *monomers*. A *polymer* of type $\sigma \in \Sigma^{(m)}$, which we also call a σ-polymer, is a string $\perp_\sigma \sigma^i$, $i \geq 0$; its *length* is i and we say that the polymer is *empty* if its length is 0. We call $\perp_\sigma$ a *stub* and let $\perp = \{\perp_\sigma \mid \sigma \in \Sigma^{(m)}\} \subset \Sigma$. Reactions can produce stubs from molecules of other species in Σ; this is an important way in which our model differs from previous work [5,12]. Polymer reactions also involve molecules in a set $\mathcal{A} = \{A_\sigma \mid \sigma \in \Sigma^{(m)}\}$ of *active query molecules*, where $\mathcal{A} \subseteq \Sigma - \Sigma^{(m)}$. For each $\sigma \in \Sigma^{(m)}$ there is a reversible polymer reaction, with the forwards and backwards directions corresponding to σ-push and σ-pop, respectively:

$$[\perp_\sigma \ldots] + \sigma \;\rightleftharpoons\; [\perp_\sigma \ldots \sigma] + A_\sigma.$$

Later, we will introduce "inactive" variants of the A_σ molecules, both to help control when pushes and pops happen, and to help track whether a polymer is empty (has length 0) or not.

Configurations. A *configuration* specifies how many molecules of each species are in the system, keeping track also of the lengths of all σ-polymers. Formally, a configuration is a mapping $\mathbf{c} : \Sigma \cup \{\perp_\sigma \sigma^i \mid i \geq 1\} \to \mathbb{N}$, where $\mathbb{N}$ is the set of nonnegative integers. We let $\mathbf{c}([\perp_\sigma \ldots])$ denote total number of σ-polymers in the system (including stubs) and let $\mathbf{c}([\perp_\sigma \ldots \sigma])$ denote total number of σ-polymers in the system that have length at least 1. With respect to configuration $\mathbf{c}$, we say that a molecule of species $\sigma \in \Sigma$ is a *leader* if $\mathbf{c}(\sigma) = 1$, and we say that a σ-polymer is a leader if $\mathbf{c}([\perp_\sigma \ldots]) = 1$.

A reaction of type (1) is *applicable* to configuration $\mathbf{c}$ if, when $r \neq r'$, $\mathbf{c}(r) \geq 1$ and $\mathbf{c}(r') \geq 1$, and when $r = r'$, $\mathbf{c}(r) \geq 2$. If the reaction is applied to $\mathbf{c}$, a new configuration $\mathbf{c}'$ is reached, in which the counts of r and r' decrease by 1 (when $r = r'$ the count of r decreases by 2), the counts of p and p' increase by 1 (when $p = p'$ the count of p increases by 2), and all other counts remain unchanged.

A σ-push is applicable if $\mathbf{c}([\perp_\sigma \ldots]) > 0$ and $\mathbf{c}(\sigma) > 0$, and a σ-pop is applicable if $\mathbf{c}([\perp_\sigma \ldots \sigma]) > 0$ and $\mathbf{c}(A_\sigma) > 0$. The *result* of applying a σ-push reaction is that $\mathbf{c}(\sigma)$ decreases by 1, $\mathbf{c}(A_\sigma)$ increases by 1 and also for exactly one

$i \geq 0$ such that $\mathbf{c}(\perp_\sigma \sigma^i) > 0$, $\mathbf{c}(\perp_\sigma \sigma^i)$ decreases by 1 and $\mathbf{c}(\perp_\sigma \sigma^{i+1})$ increases by 1. Similarly, the *result* of applying a σ-pop reaction is that $\mathbf{c}(\sigma)$ increases by 1, $\mathbf{c}(A_\sigma)$ decreases by 1, and for exactly one $i \geq 1$ such that $\mathbf{c}(\perp_\sigma \sigma^i) > 0$, $\mathbf{c}(\perp_\sigma \sigma^i)$ decreases by 1 and $\mathbf{c}(\perp_\sigma \sigma^{i-1})$ increases by 1. Intuitively, the length of one σ-polymer in the system either grows or shrinks by 1 and correspondingly the count of A_σ either increases or decreases by 1. The affected polymer is chosen nondeterministically; exactly how the polymer is chosen is not important in the context of stable computation. For example, the polymer could be chosen uniformly at random, consistent with the model of Lakin and Phillips [12]. We defer further discussion of this to Sect. 5.

If $\mathbf{c}'$ results from the application of some reaction to $\mathbf{c}$, we write $\mathbf{c} \to \mathbf{c}'$ and say that $\mathbf{c}'$ is *directly reachable* from $\mathbf{c}$. We say that $\mathbf{c}'$ is reachable from $\mathbf{c}$ if for some $k \geq 0$ and configurations $\mathbf{c}_1, \mathbf{c}_2, \ldots, \mathbf{c}_k$,

$$\mathbf{c} \to \mathbf{c}_1 \to \mathbf{c}_2 \ldots \to \mathbf{c}_k \to \mathbf{c}'.$$

Computations and Stable Computations. We're interested in CRNs that compute, starting from some initial configuration $\mathbf{c}_0$ that contains an input. For simplicity, we focus on CRNs that compute functions $f : \mathbb{N}^k \to \mathbb{N}$. For example, the function may be Square, namely $f(n) = n^2$.

In a function-computing psCRN, the input $\mathbf{n} = (n_1, \ldots, n_k) \in \mathbb{N}^k$ is represented by counts of species in a designated set $\mathcal{I} = \{X_1, X_2, \ldots, X_k\} \subseteq \Sigma^{(m)}$ and the output is represented by the count of a different designated species $Y \in \Sigma^{(m)}$. In the initial configuration $\mathbf{c}_0 = \mathbf{c}_0(n)$, the initial counts of the input species X_i is n_i, $1 \leq i \leq k$, and the counts of all species other than the input species, including polymers and active query molecules, is 0, with the following exceptions. First, there may be some leader molecules or polymers present. Second, the count of a designated "blank" species $B \in \Sigma$ may be positive. Blank molecules are useful in order to keep all reactions bimolecular, since a unimolecular reaction $r \to p$ can be replaced by $r + B \to p + B$ (if B's are guaranteed to always be present). Blanks can also be used to create new copies of a particular molecular species.

A *computation* of a psCRN is a sequence of configurations starting with an initial configuration $\mathbf{c}_0$, such that each configuration (other than the first) is directly reachable from its predecessor. Let $\mathcal{C}$ be a psCRN, and let $\mathbf{c}$ be a configuration of $\mathcal{C}$. We say that $\mathbf{c}$ is *stable* if for all configurations $\mathbf{c}'$ reachable from $\mathbf{c}$, $\mathbf{c}(Y) = \mathbf{c}'(Y)$, where Y is the output species. The psCRN *stably computes* a given function $f : \mathbb{N}^k \to \mathbb{N}$ if on any input $\mathbf{n} \in \mathbb{N}^k$, for any configuration $\mathbf{c}$ reachable from $\mathbf{c}_0(\mathbf{n})$, a stable configuration $\mathbf{c}'$ is reachable from $\mathbf{c}$ and moreover, $\mathbf{c}'(Y) = f(n)$. Finally if psCRN $\mathcal{C}$ stably computes a given predicate, we say that $\mathcal{C}$ is *committing* if $\mathcal{C}$ has a special "commit" species L_H such that for all $\mathbf{n} \in \mathbb{N}^k$, for any configuration $\mathbf{c}$ reachable from $\mathbf{c}_0(\mathbf{n})$ that contains species L_H, if $\mathbf{c}'$ is reachable from $\mathbf{c}$ then $\mathbf{c}'$ also contains L_H and $\mathbf{c}'(Y) = f(n)$.

Bottleneck Reactions. In our CRN algorithms of Sect. 3, many reactions involve a leader molecule, representing a program counter, that reacts with a leader polymer. Such reactions, in which the count of both reactants is 1, is often described as a bottleneck reaction [15]. As explained in Sect. 5, in a stochastic CRN that executes in a well-mixed system with volume V, the expected time for such a reaction is $\Theta(V)$ [6]. Our motivation for the anonymous polymer model in Sect. 4 is to explore how to compute with polymers in a way that reduces bottleneck reactions.

3 Stable, Turing-Universal Computation by Sequential PsCRNs with Leader Polymers

Here we describe how psCRNs with leader polymers can stably simulate register machines, thereby achieving Turing-universal computation. Before doing so, we first introduce psCRN "pseudocode" which is convenient for describing psCRN algorithms. Then, as an illustration, we describe a psCRN to compute the Square function $f(n) = n^2$. We first do this for a slightly different input convention than that described in Sect. 2: we assume that all input molecules are "pre-loaded" on polymers. For this pre-loaded input model, building strongly on a construction of Cummings et al. [4], we show how committing psCRNs can simulate register machines, thereby achieving Turing-universal computation. Finally, we remove the requirement that the input is pre-loaded by adding mechanisms to detect when an input is loaded, and to restart the simulator in this case.

Sequential psCRN Pseudocode. Following earlier work [4–6,16], we describe a psCRN program as a sequence of instructions, ordered consecutively starting at 1. Because one instruction must finish before moving on to the next, we call these *sequential* psCRNs. Corresponding to each instruction number i is a molecular "program counter" species $L_i \in \Sigma$. One copy of L_1 is initially present, and no other $L_{i'}$ for $i' \neq i$ is initially present.

The instructions $\texttt{inc}(\sigma)$ and $\texttt{dec}(\sigma)$ of Table 1 increase and decrease the length of a σ-polymer by 1, respectively, making it possible to use the polymers as counters. We assume that always a sufficient number of blanks are in the system in order for the $\texttt{inc}()$ instruction to proceed. In order to ensure that the push and pop reactions happen only within the $\texttt{inc}()$ and $\texttt{dec}()$ reactions, the $\texttt{inc}(\sigma)$ operation generates the active query A_σ, which is converted into an inactive variant $I_\sigma \in \Sigma - \Sigma^{(m)}$ before the instruction execution completes, and the $\texttt{dec}(\sigma)$ instruction reactivates A_σ in order to reduce the length of a σ-polymer by 1. If a psCRN executes only instructions of Table 1, starting from an initial configuration in which there is no polymer of length greater than 0, then we have the following invariant:

Invariant: Upon completion of any instruction, the count of I_σ equals the sum of the lengths of σ-polymers.

The $\texttt{jump-if-empty}$ instruction is useful when there is a leader σ-polymer. This σ-polymer is empty (has length 0) if and only if a stub $\perp_\sigma$ is in the system. Assuming that our invariant holds, the leader σ-polymer is not empty if and only if at least one I_σ molecule is in the system. Either way, the instruction ensures that the program counter advances properly. When the σ-polymer is empty, the $\text{dec}(\sigma)$ cannot proceed and causes an algorithm to stall. The $\text{jump-if-empty}(\sigma, k)$ instruction provides a way to first check whether the σ-polymer is empty, and if not, $\text{dec}(\sigma)$ can safely be used. The create and destroy instructions provide a way to create and destroy copies of a species. For clarity, we also include $\text{create-polymer}(\sigma)$ and $\text{destroy-polymer}(\sigma)$ instructions, which create and destroy the stub $\perp_\sigma$, respectively. While more than one reaction is needed to implement one instruction, all will have completed when the instruction has completed and the program counter is set to the number of the next instruction to be executed in the pseudocode.

Table 1. Instruction abstractions of psCRN reactions. The decrement $\text{dec}(\sigma)$ instruction can complete only if some σ-polymer has length is at least 1.

i: $\text{inc}(\sigma)$	$L_i + B \longrightarrow L_i^* + \sigma$
	$\sigma + [\perp_\sigma \ldots] \rightleftharpoons A_\sigma + [\perp_\sigma \ldots \sigma]$
	$L_i^* + A_\sigma \longrightarrow L_{i+1} + I_\sigma$
i: $\text{dec}(\sigma)$	$L_i + I_\sigma \longrightarrow L_i^* + A_\sigma$
	$A_\sigma + [\perp_\sigma \ldots \sigma] \rightleftharpoons \sigma + [\perp_\sigma \ldots]$
	$L_i^* + \sigma \longrightarrow L_{i+1} + B$
i: jump-if $\text{-empty}(\sigma, k)$	$L_i + \perp_\sigma \longrightarrow L_k + \perp_\sigma$
	$L_i + I_\sigma \longrightarrow L_{i+1} + I_\sigma$
i: $\text{goto}(k)$	$L_i + B \longrightarrow L_k + B$
i: $\text{create}(\sigma)$	$L_i + B \longrightarrow L_{i+1} + \sigma$
i: $\text{destroy}(\sigma)$	$L_i + \sigma \longrightarrow L_{i+1} + B$
i: $\text{create-polymer}(\sigma)$	$L_i + B \longrightarrow L_{i+1} + \perp_\sigma$
i: $\text{destroy-polymer}(\sigma)$	$L_i + \perp_\sigma \longrightarrow L_{i+1} + B$
i: halt	$L_i + B \longrightarrow L_H + B$

Pseudocode instructions may also be function calls, where a function is itself a sequence of instructions expressed as pseudocode. Suppose again that there is a leader σ-polymer and also a leader σ'-polymer in the system. Then the $\text{copy}(\sigma, \sigma')$ function (using a temporary τ-polymer) extends the length of the σ'-polymer by the length of the σ-polymer. Another useful function is $\text{flush}(\sigma)$ which decrements the (leader) σ-polymer until its length is 0. A third function, $\text{release-output}(\sigma)$, is useful to "release" molecules on a (leader) σ-polymer as Y molecules into the solution. This function uses an additional special leader Y'-polymer which is empty in the initial configuration, and whose length at the end of the function equals the number of released Y molecules. The Y' molecule will be useful later, when we address how a psCRN can be restarted (and should not be used elsewhere in the code).

i: copy(σ, σ')	i: goto$(i.1)$
	$i.1$: create-polymer(τ)
	$i.2$: jump-if-empty$(\sigma, i.7)$
	$i.3$: dec(σ)
	$i.4$: inc(σ')
	$i.5$: inc(τ)
	$i.6$: goto$(i.2)$
	$i.7$: jump-if-empty$(\tau, i.11)$
	$i.8$: dec(τ)
	$i.9$: inc(σ)
	$i.10$: goto$(i.7)$
	$i.11$: destroy-polymer(τ)
	$i.12$: goto$(i+1)$

i: flush(σ)	i: goto$(i.1)$
	$i.1$: jump-if-empty$(\sigma, i+1)$
	$i.2$: dec(σ)
	$i.3$: goto$(i.1)$

i: release-output(σ)	i: goto$(i.1)$
	$i.1$ jump-if-empty$(\sigma, i+1)$
	$i.2$ dec(σ)
	$i.3$ inc(Y')
	$i.4$ create(Y)
	$i.5$ goto$(i.1)$

Numbering of Function Instructions. For clarity, we use $i.1, i.2$, and so on to label the lines of a function called from line i of the main program. Upon such a function call, the CRN's program counter first changes from L_i to $L_{i.1}$. The program counter is restored to L_{i+1} upon completion of the function's instructions, e.g., via a goto$(i+1)$ instruction or a jump-if-empty$(\sigma, i+1)$ instruction. If one function f_B is called from line $a.b$ of another function f_A, the program counter labels would be $a.b.1$, $a.b.2$ and so on, and so the label "i" in the function description should be interpreted as "$a.b$". In this case, when the function f_B completes, control is passed back to line $a.(b+1)$ of function f_A; that is, the "goto$(i+1)$" statement should be interpreted as "goto$(a.(b+1))$". Also for clarify, we use special labeling of instructions in a few special places, such as the restart function below, in which instructions are labeled $s1, s2$ and so on.

psCRNs with Pre-loaded Inputs. As noted in the introduction, a challenge in achieving stable computation with psCRNs is detecting the absence of inputs. To build up to our methods for addressing this challenge, we first work with a more convenient convention, that of pre-loaded inputs. By this we mean that if the input contains n_i molecules of a given species X_i, then in the initial configuration there is a unique X_i-polymer of length n_i (the "pre-loaded" polymer). Furthermore, there are n_i copies of the inactive query molecule I_{X_i} in the system. Intuitively, the pre-loaded initial configuration is one that would be reached if n_i inc(X_i) operations were performed from an initial configuration with no inputs and an empty X_i-polymer, for every input species X_i.

A Committing, Sequential psCRN with Pre-loaded Inputs for Square. Our psCRN for the Square function $f(n) = n^2$ has one input species X and one output species Y. In the pre-loaded initial configuration, the input is represented as the length n of a leader X-polymer, and the count of I_X is n. The number of blanks in the initial configuration must be greater than n^2, since blanks are used to produce the n^2 output molecules. The only other molecule in the initial configuration is the leader program counter L_1. The psCRN has a loop (implemented using jump-if-empty and goto) that executes n times, adding n to an intermediate Y_{int}-polymer each time. When the loop completes, the output is released from the Y_{int}-polymer in the form of Y, so that the number of Y's in solution is n^2, and the psCRN halts. The halting state is in effect a committing state, since no transition is possible from L_H.

Algorithm 1. Sequential-n^2-psCRN, with input n pre-loaded on X-polymer.

```
1: create-polymer(X')
2: create-polymer(Y_int)
3: copy(X, X')
4: jump-if-empty(X',8)
5:     dec(X')
6:     copy(X, Y_int)
7:     goto(4)
8: release-output(Y_int)
9: halt
```

Committing Turing-Universal Computation by psCRNs with Pre-loaded Inputs. Turing-universal computation is possible with register machines (RMs). To simulate a halting register machine that computes function $f : \mathbb{N}^k \to \mathbb{N}$ with $r \geq k$ unary registers, a psCRN has r unary counters $R_1, R_2, \ldots, R_r$, the first k of which initially contain the input counts $n_1, n_2, \ldots, n_k$, while the others are initially 0. Throughout the simulation of the register machine, the psCRN has exactly one R_l-polymer for each register R_l, $1 \leq l \leq r$. In addition, there is one additional polymer, a Y'-polymer, which is initially empty and is used by the release-output function. A register machine program is a sequence of instructions, where instructions can increment a register; decrement a non-empty register; test if a register is empty (0) and jump to a new instruction if so; or halt. Table 1 already shows how all four of these instructions can be implemented using a psCRN. We assume in what follows that these are the only instructions used by the psCRN simulator; in particular, no additional registers (polymers) are ever created or destroyed. If register R_r is used to store the output, then the output is released into solution, using release-output(R_r), once the machine being simulated reaches its halt state. We assume that release-output(R_r) is the only function call of the RM simulator.

Stable, Turing-Universal Computation by psCRNs. We now handle the case that the input is represented as counts of molecules, rather than pre-loaded polymers. That is, in the initial configuration of the psCRN all polymers of types $R_1, R_2, \ldots, R_r$ are empty, and instead, for each input n_l, $1 \leq l \leq k$, there are n_l copies of molecule R_l in solution. Our scheme uses the R_l-push reaction to load inputs. We add CRNs to detect input-loading and to restart the simulator in this case. Once all inputs are loaded, the system is never subsequently restarted. Overall our simulation has four components:

- **Input loading:** This is done as R_l-push, which can happen at any time until all inputs are loaded. Recall that the R_l-push reactions are
$$[\perp_{\mathbf{R}_l} \ldots] + \mathbf{R}_l \longrightarrow [\perp_{\mathbf{R}_l} \ldots \mathbf{R}_l] + A_{R_l}, 1 \leq l \leq k.$$
Each such reaction generates an active query molecule A_{R_l} which, as explained below, triggers input detection.
- **Register machine (RM) simulation:** Algorithm 2 shows the simulator in the case of three input registers. This psCRN program has a "prelude" phase that starts by creating three new polymers R_1', R_2' and R_3' (lines P1, P2, and P3), and then copies the input register polymers R_1, R_2, and R_3 to polymers R_1', R_2', and R_3', respectively (lines P4, P5, and P6). Then starting from line numbered 1, the simulation uses register R_l' rather than R_l, $1 \leq l \leq 3$, as well as the remaining initially empty registers $R_4, \ldots, R_r$. Upon completion of the computation, the output is released from register R_r, and the simulator halts (produces the L_H species).

Algorithm 2. Sequential-RM-psCRN, $k = 3$ input registers, r registers in total.

P1: `create-polymer(`R_1'`)`
P2: `create-polymer(`R_2'`)`
P3: `create-polymer(`R_3'`)`
P4: `copy(`R_1, R_1'`)`
P5: `copy(`R_2, R_2'`)`
P6: `copy(`R_3, R_3'`)`
1: // Rest of psCRN simulation pseudocode here, with
2: // R_1, R_2 and R_3 replaced by R_1', R_2' and R_3'
    ⋮   ...
   : // ending with `release-output(`R_r`)` function and `halt` instruction.

- **Input detection:** This is triggered by the presence of an active query molecule A_{R_l}. For each value i of the main program counter after the prelude phase (i.e., after lines P1 through P6) and for L_H we have the following reactions, where s1 is the first number of the **restart** pseudocode (see below). The reactions convert the active A_{R_l} molecule into its inactive counterpart, I_{R_l}, since the input molecule is now loaded, and also changes the program counter to L_{s1}, which triggers restart.

$$L_i + A_{R_l} \longrightarrow L_{s1} + I_{R_l}, 1 \leq l \leq k,$$
$$L_H + A_{R_l} \longrightarrow L_{s1} + I_{R_l}, 1 \leq l \leq k.$$

L_H is no longer a committing species, since it may change to L_{s1}.

No input detection is done in prelude lines P1, P2, and P3. Line P4 is a function call to $\mathtt{copy}(R_1, R_1')$, which executes instructions numbered P4.1 through P4.11 of $\mathtt{copy}$. Input detection is only done at line P4.2, the first $\mathtt{jump\text{-}if\text{-}empty}$ instruction:

$$L_{P4.2} + A_{R_1} \longrightarrow L_{P4.2} + I_{R_1}.$$

It does not trigger a restart, but simply converts the active query molecule A_{R_1} to I_{R_1}. Similarly, for lines P5 and P6, we add the reactions

$$L_{P5.2} + A_{R_2} \longrightarrow L_{P5.2} + I_{R_2} \text{ and}$$
$$L_{P6.2} + A_{R_3} \longrightarrow L_{P6.2} + I_{R_3}.$$

- **Restart:** Restart happens a number of times that is at most the total input length $n_1 + n_2 + \ldots n_k$, since each input molecule is loaded into a register exactly once, generating one active query molecule. For $k = 3$, the registers R_1', R_2' and R_3', as well as the registers $R_4, \ldots, R_r$ are flushed, and any outputs that have been released in solution are destroyed, assuming that the number of outputs released into the solution was tracked by some Y' register, as before. Then the program counter is set to line P4 of the simulator (leader molecule L_{P4}). Algorithm 3 shows the restart pseudocode.

Algorithm 3. Restart

---
s1: $\mathtt{flush}(R_1')$
s2: $\mathtt{flush}(R_2')$
s3: $\mathtt{flush}(R_3')$
s4: $\mathtt{flush}(R_4)$
$\quad \ldots$
sr: $\mathtt{flush}(R_r)$
s($r+1$): $\mathtt{destroy\text{-}output}()$
s($r+2$): $\mathtt{goto}(\text{P4})$

---

i: $\mathtt{destroy\text{-}output}()$	i: $\mathtt{goto}(i.1)$
	$i.1$: $\mathtt{jump\text{-}if\text{-}empty}(Y', i.5)$
	$i.2$: $\quad \mathtt{dec}(Y')$
	$i.3$: $\quad \mathtt{destroy}(Y)$
	$i.4$: $\quad \mathtt{goto}(i.1)$
	$i.5$: $\mathtt{goto}(i+1)$

Correctness of Algorithm 2: Sequential-RM-psCRN. We claim that our register machine simulator without pre-loaded inputs stably computes the same function as the register machine, assuming that sufficiently many blank molecules B are present to ensure that reactions with B as a reactant can always proceed. (We note that no "fairness" assumption regarding the order in which reactions happen is necessary to show stability, since stability is a "reachability" requirement.)

The first three instructions, lines P1, P2, and P3), simply create polymers R_1', R_2', and R_3'. The delicate part of the simulation lies in the next three instructions on lines P4, P5, and P6, which copy input registers R_1, R_2, and R_3 to R_1', R_2' and R_3', respectively. For concreteness, consider the instruction $\texttt{copy}(R_1, R_1')$ in line P4. (The argument is the same for line P5, with R_2, R_2' substituted for R_1, R_1', and is also the same for line P6, with R_3, R_3' substituted for R_1, R_1'.)

The R_1-push reaction used in input loading can cause a violation of our earlier invariant that upon completion of each instruction, the count of I_{R_1} equals the length of the (leader) R_1-polymer. Instead, we have that I_{R_1} is less than or equal to the length of the R_1-polymer. This can cause the $\texttt{jump-if-empty}$ instruction numbered P4.2 in the $\texttt{copy}$ function to stall, when there is no I_{R_1} and also no $\perp_{R_1}$ (since the R_1-polymer is not empty due to input loading). In this case, the input detection reaction (introduced above)

$$L_{\text{P4.2}} + A_{R_1} \longrightarrow L_{\text{P4.2}} + I_{R_1},$$

will convert A_{R_1} to I_{R_1}. This averts stalling, since $\texttt{jump-if-empty}$ can proceed using I_{R_1}. The subsequent lines of the $\texttt{copy}(R_1, R_1')$ code can then proceed.

Once lines P4, P5, and P6 have completed, the correctness of the psCRN simulation of the register machine, using the copies R_1', R_2' and R_3', is not affected by input loading. Input loading and input detection can also proceed. These are the only viable reactions from the "halting" state L_H, and so eventually (since the RM machine being simulated is a halting machine), on any sufficiently long computation path, all inputs must be loaded and detected. Input detection after the prelude phase produces a "missing" I_σ molecule and triggers a restart. The restart flushes all registers used by the simulator, and also, using the Y-polymer, destroys any outputs in solution. (Since restart is triggered only when the program counter is at a line of the main program, restart does not interrupt execution of the $\texttt{release-output}$ function.) A new simulation is then started at line P4, ensuring that any inputs that have been loaded since the last detect are copied to the simulator's input registers R_1', R_2' and R_3'.

Once all inputs have been detected, the invariant is restored and the simulator proceeds correctly, producing the correct output. This correct output is never subsequently changed, and so the computation is stable.

Bottleneck Reactions. In our sequential psCRNs, both $\texttt{inc}(\sigma)$ and $\texttt{dec}(\sigma)$ contain bottleneck reactions, and so $\texttt{copy}(\sigma, \sigma')$ has $\Theta(|\sigma|)$ bottleneck reactions. Thus the psCRN for Square has $\Theta(n^2)$ bottleneck reactions. In the next section we show how to compute a close variant of the Square function with fewer bottleneck reactions, using anonymous polymers rather than leader polymers.

4   Faster Computation of Square by Threaded psCRNs with Anonymous Polymers

To avoid the bottleneck reactions of our sequential psCRN, we enable the $\texttt{inc}(\sigma)$ and $\texttt{dec}(\sigma)$ instructions to operate on many functionally-identical "anonymous" σ-polymers, rather than just a single leader polymer. Here we describe how this can work, using the function $f(n) = n2^{\lfloor \lg n \rfloor}$ as an example, and focusing only on the pre-loaded input model. Error detection and correction can be layered on, in a manner similar to Sect. 3.

We start with a single Y-polymer of length n, and we wish to create a total of $2^{\lfloor \lg n \rfloor}$ Y-polymers, whose lengths sum to $n2^{\lfloor \lg n \rfloor}$. Algorithm 4 proceeds in $\lfloor \lg n \rfloor$ rounds (lines 6–9), doubling the number of Y's on each round. To keep track of the Y molecules, we introduce a distributed σ-counter data structure, and use it with $\sigma = Y$. The data structure consists of σ-polymers that we call σ-thread-polymers, plus a thread-polymer counter T_σ, which is a leader polymer whose length, $|T_\sigma|$, is the number of σ-thread-polymers. The *value* of this distributed counter is the total length of all σ-thread-polymers. We explain below how operations on this distributed counter work.

Algorithm 4. Threaded-$n2^{\lfloor logn \rfloor}$-psCRN.

```
 1: create-polymer(H)
 2: copy(X, H)
 3: create-distributed-counter(Y)
 4: add-thread-polymer(Y,1)
 5: copy(X, Y)
 6: halve(H)
 7: jump-if-empty(H,10)
 8:     double(Y)
 9:     goto(6)
10: halt
```

Algorithm 4 counts the number of rounds using a leader H-polymer, whose length is halved on each round. The $\texttt{halve}$ function is fairly straightforward to implement, using the instructions and functions already introduced in Sect. 3.

i: halve(H)	i: goto($i.1$)
	$i.1$: create-polymer(H')
	$i.2$: copy(H, H')
	$i.3$: jump-if-empty(H', $i.9$)
	$i.4$: dec(H')
	$i.5$: dec(H)
	$i.6$: jump-if-empty(H', $i.9$)
	$i.7$: dec(H')
	$i.8$: goto($i.3$)
	$i.9$: destroy-polymer(H')
	$i.10$: goto($i + 1$)

The double(σ) function of Algorithm 4 is where we leverage our distributed Y-counter (with $\sigma = Y$). Recall that a distributed σ-counter data structure consists of a set of anonymous σ-polymers, which we call σ-thread-polymers, plus a thread-polymer counter T_σ, which is a leader polymer whose length is the number of σ-thread-polymers. The double function first creates two other distributed counters τ and τ' (lines $i.1$ and $i.2$), and gives each the same number of thread-polymers as σ, namely $|T_\sigma|$ thread-polymers (lines $i.3$ and $i.4$), all of which are empty. The heart of double (line $i.5$) transfers the contents of the distributed σ-counter to τ and τ', emptying and destroying all σ-thread-polymers in the process. It then creates double the original number of (empty) σ-thread-polymers (lines $i.6$ and $i.7$; note that the number of threads of τ is the original value of $|T_\sigma|$). It finally transfers the τ and τ' polymers back to σ (lines $i.8$ and $i.9$), thereby doubling σ.

i: double(σ)	$i.1$ create-distributed-counter(τ)
	$i.2$ create-distributed-counter(τ')
	$i.3$ add-thread-polymers(τ, T_σ)
	$i.4$ add-thread-polymers(τ', T_σ)
	$i.5$ transfer(σ, τ, τ')
	$i.6$ add-thread-polymers(σ, T_τ)
	$i.7$ add-thread-polymers($\sigma, T_{\tau'}$)
	$i.8$ transfer(τ, σ)
	$i.9$ transfer(τ', σ)
	$i.10$ destroy-distributed-counter(τ)
	$i.11$ destroy-distributed-counter(τ')
	$i.12$ goto($i+1$)

Next are details of instructions used to create an empty distributed counter, and to add empty threads to the counter. Again, these are all straightforward sequential implementations (no threads), using leader polymers to keep track of counts.

i: create-distributed-counter(σ)	i: goto($i.1$)
// Creates an empty counter	$i.1$ create-polymer(T_σ)
// with zero polymers	$i.2$ goto($i+1$)
i add-thread-polymers(σ, T)	i goto($i.1$)
// Adds $|T|$ empty	$i.1$: create-polymer(Temp)
// thread-polymers to the	$i.2$: copy(T,Temp)
// distributed σ-counter,	$i.3$: jump-if-empty(Temp, $i.8$)
// where T is a counter	$i.4$: dec(Temp)
	$i.5$: create-polymer(σ)
	$i.6$: inc(T_σ)
	$i.7$: goto($i.3$)
	$i.8$: destroy-polymer(Temp)
	$i.9$: goto($i+1$)
i add-thread-polymer($\sigma, 1$)	i goto($i.1$)
// Adds one empty	$i.1$: create-polymer(σ)
// thread-polymer to the	$i.2$: inc(T_σ)
// distributed σ-counter	$i.3$: goto($i+1$)

The **transfer** function transfers the value of a distributed σ-counter to two other distributed counters called τ and τ'. In line $i.2$ of **transfer**, function **create-threads** creates T_σ identical "thread" program counters, L_t. Once again this is straightforward, using a leader polymer to keep track of counts. All of the thread program counters execute the **thread-transfer** function in line $i.4$ of **transfer**, thereby reducing bottleneck reactions (details below). The "main" program counter, now at line $i.4$ of the **transfer** function, can detect when all threads have completed, because each decrements Thread-Count exactly once, and so Thread-Count has length zero exactly when all threads have completed. At that point, the main program counter progresses to line $i.5$, destroying the thread program counters using the **destroy-threads** function (not shown, but uses the **destroy** function to destroy each single thread).

i: $\mathtt{transfer}(\sigma,\tau,\tau')$	i: $\mathtt{goto(i.1)}$
// transfer σ to	$i.1$: $\mathtt{create\text{-}polymer}$(Thread-Count)
// both τ and τ'	$i.2$: $\mathtt{create\text{-}threads}(T_\sigma,\, L_t)$
	$i.3$: $\mathtt{copy}(T_\sigma,$ Thread-Count$)$
	$i.4$: $\mathtt{loop\text{-}until\text{-}empty}$(Thread-Count, $i.5$)
	$\qquad$ $\mathtt{thread\text{-}transfer}(\sigma,\tau,\tau',$Thread-Count$)$
	$i.5$: $\mathtt{destroy\text{-}threads}(L_t)$
	$i.6$: $\mathtt{destroy\text{-}polymer}$(Thread-Count)
	$i.7$: $\mathtt{goto}(i+1)$

The function $\mathtt{transfer}(\sigma, \tau)$, not shown but used in **double**, is the same as $\mathtt{transfer}(\sigma, \tau, \tau')$, except the call to **thread-transfer** does not include τ' and the "inc(τ')" line is removed in the implementation of **thread-transfer**.

i: $\mathtt{create\text{-}threads}(T_\sigma, L_t)$	i: $\mathtt{goto(i.1)}$
// create T_σ thread	$i.1$: $\mathtt{create\text{-}polymer}$(Temp)
// program counters, L_t	$i.2$: $\mathtt{copy}(T_\sigma,$ Temp$)$
	$i.3$: $\mathtt{jump\text{-}if\text{-}empty}$(Temp, $i.7$)
	$i.4$ $\quad$ $\mathtt{dec}$(Temp)
	$i.5$ $\quad$ $\mathtt{create}(L_t)$
	$i.6$ $\quad$ $\mathtt{goto}(i.3)$
	$i.7$: $\mathtt{destroy\text{-}polymer}$(Temp)
	$i.8$: $\mathtt{goto}(i+1)$
i : $\mathtt{loop\text{-}until\text{-}empty}(\sigma, k)$	$L_i + \perp_\sigma \longrightarrow L_k + \perp_\sigma$

Finally, we describe how threads work in **thread-transfer**. The threadon() function executes $|T_\sigma|$ times, one per copy of L_t, thereby creating $|T_\sigma|$ L_{t_1} program counters that execute computation "threads". Using the function **dec-until-destroy-polymer**, each thread repeatedly (zero or more times) decrements one of the σ-thread-polymers and then increments both τ and τ'. This continues until the thread finds an empty σ-thread-polymer, i.e., the stub $\perp_\sigma$, in which case it destroys the stub and moves to line $t.5$. The dec(σ) and inc(σ) functions of Sect. 3 work exactly as specified, even when applied to distributed counters. A key point is that the threads work "anonymously" with the thread-polymers; it is not the case that each thread "owns" a single thread-polymer. Accordingly, one thread may do more work than another, but in the end all thread-polymers are empty.

A thread exits the `dec-until-destroy-polymer` loop by destroying exactly one σ-polymer. Since at the start of `thread-transfer` the number of σ-thread-polymers equals the number of thread program counters, all thread program counters eventually reach line $t.5$, and there are no σ-thread-polymers once all threads have reached line $t.5$ of the code. At line $t.5$, each thread decrements Thread-Count, and then stalls at line $t.6$. Moreover, once all threads have reached line $t.6$, polymer ThreadCount is empty. At this point, the program counter for `transfer` changes from line $i.4$ to line $i.5$, and all thread program counters are destroyed.

i: thread-transfer $(\sigma,\tau,\tau',\text{Thread-Count})$	i: threadon()
	$t.1$: dec-until-destroy-polymer$(\sigma, t.5)$
	$t.2$: inc(τ)
	$t.3$: inc(τ')
	$t.4$: goto$(t.1)$
	$t.5$: dec(Thread-Count)
	$t.6$:

i: **threadon()**:	$L_i + L_t \longrightarrow L_i + L_{t.1}$
i: **dec-until-destroy** **-polymer**(σ, k)	$L_i + I_\sigma \longrightarrow L_i^* + A_\sigma$
	$A_\sigma + [\bot_\sigma \ldots \sigma] \rightleftharpoons \sigma + [\bot_\sigma \ldots]$
	$L_i^* + \sigma \longrightarrow L_{i+1} + \mathrm{B}$
	$L_i + \bot_\sigma \longrightarrow L_i^{**} + \mathrm{B}$
	$L_i^{**} + I_{T_\sigma} \longrightarrow L_i^{***} + A_{T_\sigma}$
	$A_{T_\sigma} + [\bot_{T_\sigma} \ldots T_\sigma] \rightleftharpoons T_\sigma + [\bot_{T_\sigma} \ldots]$
	$L_i^{***} + T_\sigma \longrightarrow L_k + \mathrm{B}$

Correctness. We claim that on any input $n \geq 0$, pre-loaded on a leader X-polymer, Algorithm 4: Threaded-$n2^{\lfloor \lg n \rfloor}$-psCRN eventually halts with the value of the distributed-Y-counter being $f(n) = n2^{\lfloor \log n \rfloor}$.

The algorithm creates and initializes H to be a polymer of length n (lines 1–2), and the Y-distributed-counter to have a single polymer-thread of length n (lines 3–5). When $n = 0$, H is empty, so from line 6 the algorithm jumps to line 10 and halts, with the value of Y being $f(0) = 0$ as claimed.

Suppose that $n > 0$. Reasoning about the `halve` function is straightforward, since it is fully sequential. We claim that in each round of the algorithm (lines 6–9), lines 7 and 8 complete successfully, with $|H|$ halving (that is, $|H| \to \lfloor |H|/2 \rfloor$) in line 7, and with both the value of Y and $|T_Y|$, the number of Y-thread-polymers, doubling in line 8. As a result, $|H| = 0$ after $\lfloor \lg n \rfloor$ rounds and the algorithm halts with value$(Y) = f(n)$.

Correctness of the double function is also straightforward to show, if we show that the transfer(σ, τ, τ') (and the transfer(σ, τ) variant) works correctly.

Line $i.4$ is the core of transfer. We show that line $i.4$ does complete, that is, Thread-Count does become empty, that execution of line $i.4$ increases the values of distributed counters τ and τ' by the value of σ (while leaving the number of τ- and τ'-thread-polymers unchanged), and also changes value(σ) and the number of σ-thread-polymers to 0.

The loop-if-empty instruction ensures that the main program counter must stay at line $i.4$ of function transfer until Thread-Count is empty. Meanwhile, this main program counter can also activate threads using the threadon() function, that is, change the thread program counters from L_t to $L_{t.1}$. From line $i.2$ of transfer, the number of such thread program counters is $|T_\sigma|$.

Each of these program counters independently executes thread-transfer. At line $t.1$, either (i) a dec(σ) is performed (first three reactions of dec-until-destroy-polymer), or (ii) a σ-polymer-thread is destroyed and the polymer-thread-count T_σ is decremented (last four reactions). In case (i), both τ and τ' are incremented (lines $t.2$ and $t.3$), and the thread goes back to the dec-until-destroy-polymer instruction. In case (ii), the thread moves to line $t.4$, decrements Thread-Count exactly once, and moves to line $t.5$.

Because the number of threads equals the value of Thread-Count at the start of the loop-until-empty (line $i.4$), and because the main program counter can't proceed beyond line $i.4$ of the transfer function until Thread-Count is zero, all threads must eventually be turned on each of these threads must reach line $t.4$ and must decrement Thread-Count. Only then can the main program counter proceed to line $i.5$ of transfer. This in turn means that each thread must destroy a σ-polymer-thread. Since the number of σ-polymer-threads, $|T_\sigma|$, equals Thread-Count, all threads are destroyed (and the T_σ-polymer is empty) upon completion of thread-transfer.

Bottleneck Reactions. In each round, the halve(σ) function decreases the length of the H-polymer by a factor of 2, starting from n initially. Each decrement or increment of the H-polymer includes a bottleneck reaction, so there are $\Theta(n)$ bottleneck reactions in total, over all rounds. The double function creates 2^l thread-polymers in round l, for a total of $\Theta(n)$ thread-polymers over all rounds. The transfer function creates 2^l threads in round l and similarly destroys 2^l threads, and copies a polymer of length 2^l, so again has $\Theta(n)$ bottleneck reactions over all rounds. The reactions in thread-transfer are not bottleneck reactions (except in round 1); we analyze these in the next section.

5   psCRN Time Complexity Analysis and Simulation

We follow the stochastic model of Soloveichik et al. [6] for well-mixed, closed systems with fixed volume V. We assume that all reactions have rate constant 1. When in configuration $\mathbf{c}$, the *propensity* of reaction $R : r + r' \to p + p'$ is $\mathbf{c}(r)\mathbf{c}(r')/V$ if $r \neq r'$, and is $\binom{\mathbf{c}(r)}{2}/V$ if $r = r'$. Let $\Delta(\mathbf{c})$ be the sum of all reaction propensities, when in configuration $\mathbf{c}$. When a reaction occurs in configuration $\mathbf{c}$, the probability that it is reaction R is the propensity of R divided by $\Delta(\mathbf{c})$, and the expected time for a reaction is $1/\Delta(\mathbf{c})$. When the only applicable reaction is a bottleneck reaction, and the volume V is $\Theta(n^2)$, the expected time for this bottleneck reaction is $\Theta(n^2)$. Soloveichik et al. [6] consider CRNs without polymers, but the same stochastic model is used by Lakin et al. [12] and Qian et al. [5], where the reactants r or r' (as well as the products) may be polymers.

Expected time complexity of Algorithm 1: Sequential-n^2-psCRN. This psCRN has n rounds, with $\Theta(n)$ instructions per round; for example, the copy of length n in each round has n `inc` instructions. So the total number of instructions executed, over all rounds is $\Theta(n^2)$; moreover, there are $\Theta(n^2)$ $inc(\sigma)$ instruction overall. The program's instructions execute sequentially, that is, the ith instruction completes before the $(i+1)$st instruction starts, so the total expected time is the sum of the expected times of the individual instructions. Each instruction involves a constant number of reactions. Some instructions involve bottleneck reactions; for example, the push reaction of the `inc` instruction is a bottleneck reaction. So an execution of the program involves $\Theta(n^2)$ bottleneck reactions. Each of these takes $\Theta(n^2)$ time, so the overall expected time is $\Theta(n^4)$.

Expected time complexity of Algorithm 4: Threaded-$n2^{\lfloor logn \rfloor}$-psCRN. We noted earlier that Algorithm 4 has $\Theta(n)$ non-threaded instructions, and in fact $\Theta(n)$ bottleneck instructions. These take expected time $\Theta(n^3)$ overall, since the time for each is $\Theta(V) = \Theta(n^2)$.

Now, consider the threaded function, `thread-transfer`. In round $l, 1 \leq l \leq \lfloor \lg n \rfloor$, `thread-transfer` has 2^l threads, and pushes $n2^l$ Y monomers on to 2^l anonymous Y-polymers. Since each Y-push reaction is independent and is equally likely to increment each of the 2^l Y-polymers, the expected number of molecules per polymer is n. Using a Chernoff tail bound, we can show that all polymers have length in the range $[n/2, 2n]$ with all but negligibly small probability. In what follows, we assume that this is the case.

During the first $\geq n/2$ of the `thread-transfer` decrements in round l, the count of each of the reactants is 2^l: one program counter per thread and 2^l polymers in total. So the expected time for these decrements is $\Theta(V/2^{2l})$. Pessimistically, if all of the decrements happen to the same polymer, whose length could be as little as $n/2$ by our assumption above, there are $2^l - 1$ polymers and threads available for the next decrements, $2^l - 2$ polymers and threads available for the next n decrements after that once a second polymer is depleted, and so on. So the total expected time is $O(Vn \sum_{j=1}^{2^l} (1/j^2)) = O(nV)$. Multiplying by $\lfloor \lg n \rfloor$, the number of rounds, and noting that $V = O(n^2)$, we have that the total expected time for the `thread-transfer` over all rounds is $O(n^3 \lg n)$.

Simulator. To test the correctness of our protocols, we developed a custom CRN simulator designed to support anonymous polymers, though we only show the results of our sequential protocol here in this paper. The simulator uses a slightly modified version of Gibson and Bruck's next reaction method [17], which itself is an extension of Gillespie's algorithm [18]. We redefine what a single "species"

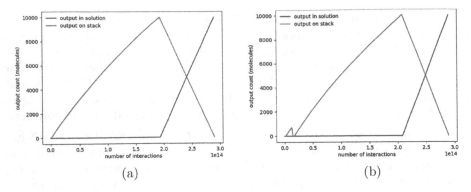

(a)   (b)

Fig. 1. (a) Simulation of $f(n) = n^2$ starting with pre-loaded input, $n = 100$. (b) Simulation of $f(n) = n^2$ with input detection and restart, $n = 100$. Each coloured line in the plots shows the count of the outputs as a function of the number of interactions, with the blue line being the count of output species Y finally released into solution, while the red line shows the size of the Y_{int} polymer. By interaction, we mean a collision of two molecular species in the system, which may or may not result in a reaction. (Color figure online)

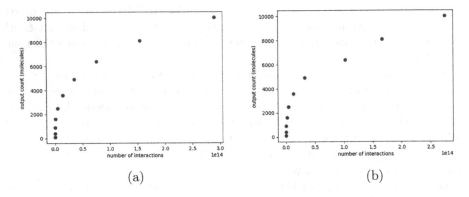

(a)   (b)

Fig. 2. (a) Simulations of $f(n) = n^2$ starting with pre-loaded input, covering a range of n. (b) Simulations of $f(n) = n^2$ with input detection and restart, covering a range of n. Each blue circle plots a single simulation, showing the final count of the output species Y against the number of interactions it took to complete. (Color figure online)

is from the algorithm's point of view, classifying all σ-polymers as one species, and track polymer lengths separately.

Interestingly, simulation of our stable, error-corrected sequential psCRN for Square usually takes little extra time compared to the committing, pre-loaded sequential psCRN (see both Figs. 1 and 2). This is because each of the n error detection steps, and subsequent restart, is expected to happen in $O(n^2)$ time, which is negligible compared to the time for the $\Theta(n^4)$ expected running time of the psCRN with fully loaded input.

6   Conclusions and Future Work

In this work, we've expanded the computing model of stochastic chemical reaction networks with polymers, by considering inputs that are represented as monomers in solution, as well as anonymous polymers that facilitate distributed data structures and threaded computation. We've shown that stable, error-free Turing-universal computation is possible in the monomer input model, by introducing an error-correction scheme that takes advantage of the ability to check for empty polymers. We've illustrated how programming with anonymous polymers can provide speed-ups, compared with using leader polymers only, and how leader polymers can be used for synchronization purposes by CRNs with anonymous polymers.

There are many interesting directions for future work. First, we have shown how to use anonymous polymers to get a speed-up for the Square problem, but we have not shown that such a speed-up is not possible without the use of anonymous polymers. Is it possible to show lower bounds on the time complexity of problems when only leader polymers are available? Or, could bottleneck reactions be reduced or avoided by a psCRN computing Square? Second, our faster psCRN for Square with anonymous polymers still uses leader polymers for synchronization. Is the speed-up possible even without the use of leader polymers? More generally, how can synchronization be achieved in leaderless psCRNs? Are there faster psCRNs, with or without leader polymers? It would be very interesting to know what problems have stable psCRNS that use no leaders, but can use anonymous polymers. Finally, it would be valuable to have more realistic models of reaction propensities for psCRN models.

References

1. Soloveichik, D., Cook, M., Winfree, E., Bruck, J.: Computation with finite stochastic chemical reaction networks. Nat. Comput. **7**(4), 615–633 (2008)
2. Chen, H.-L., Doty, D., Soloveichik, D.: Deterministic function computation with chemical reaction networks. Nat. Comput. **13**, 517–534 (2014)
3. Angluin, D., Aspnes, J., Eisenstat, D.: Stably computable predicates are semilinear. In: Proceedings of the Twenty-Fifth Annual ACM Symposium on Principles of Distributed Computing, PODC 2006, New York, pp. 292–299. ACM Press (2006)
4. Cummings, R., Doty, D., Soloveichik, D.: Probability 1 computation with chemical reaction networks. Nat. Comput. **15**(2), 245–261 (2014)
5. Qian, L., Soloveichik, D., Winfree, E.: Efficient turing-universal computation with DNA polymers. In: Sakakibara, Y., Mi, Y. (eds.) DNA 2010. LNCS, vol. 6518, pp. 123–140. Springer, Heidelberg (2011). https://doi.org/10.1007/978-3-642-18305-8_12
6. Soloveichik, D., Cook, M., Winfree, E., Bruck, J.: Computation with finite stochastic chemical reaction networks. Nat. Comput. **7**, 615–633 (2008)
7. Bennett, C.: Logical reversibility of computation. IBM J. Res. Dev. **17**(6), 525–532 (1973)
8. Bennett, C.: The thermodynamics of computation - a review. Int. J. Theor. Phys. **21**(12), 905–940 (1981)

9. Johnson, R., Winfree, E.: Verifying polymer reaction networks using bisimulation (2014)
10. Cardelli, L., Zavattaro, G.: Turing universality of the biochemical ground form. Math. Struct. Comput. Sci. **20**, 45–73 (2010)
11. Jiang, H., Riedel, M., Parhi, K.: Synchronous sequential computation with molecular reactions. In: Proceedings of the 48th Design Automation Conference, DAC 2011, New York, pp. 836–841. ACM (2011)
12. Lakin, M.R., Phillips, A.: Modelling, simulating and verifying turing-powerful strand displacement systems. In: Cardelli, L., Shih, W. (eds.) DNA 2011. LNCS, vol. 6937, pp. 130–144. Springer, Heidelberg (2011). https://doi.org/10.1007/978-3-642-23638-9_12
13. Angluin, D., Aspnes, J., Diamadi, Z., Fischer, M.J., Peralta, R.: Computation in networks of passively mobile finite-state sensors. Distrib. Comput. **18**, 235–253 (2006)
14. Chatzigiannakis, I., Michail, O., Nikolaou, S., Pavlogiannis, A., Spirakis, P.G.: Passively mobile communicating machines that use restricted space. In: Proceedings of the 7th ACM ACM SIGACT/SIGMOBILE International Workshop on Foundations of Mobile Computing, FOMC 2011, New York, pp. 6–15. ACM (2011)
15. Chen, H.-L., Cummings, R., Doty, D., Soloveichik, D.: Speed faults in computation by chemical reaction networks. In: Kuhn, F. (ed.) DISC 2014. LNCS, vol. 8784, pp. 16–30. Springer, Heidelberg (2014). https://doi.org/10.1007/978-3-662-45174-8_2
16. Angluin, D., Aspnes, J., Eisenstat, D.: Fast computation by population protocols with a leader. In: Dolev, S. (ed.) DISC 2006. LNCS, vol. 4167, pp. 61–75. Springer, Heidelberg (2006). https://doi.org/10.1007/11864219_5
17. Gibson, M.A., Bruck, J.: Efficient exact stochastic simulation of chemical systems with many species and many channels. J. Phys. Chem. A **104**(9), 1876–1889 (2000)
18. Gillespie, D.T.: Exact stochastic simulation of coupled chemical reactions. J. Phys. Chem. **81**(25), 2340–2361 (1977)

SIMD||DNA: Single Instruction, Multiple Data Computation with DNA Strand Displacement Cascades

Boya Wang[(✉)], Cameron Chalk, and David Soloveichik

The University of Texas at Austin, Austin, USA
bywang@utexas.edu

Abstract. Typical DNA storage schemes do not allow in-memory computation, and instead transformation of the stored data requires DNA sequencing, electronic computation of the transformation, followed by synthesizing new DNA. In contrast we propose a model of in-memory computation that avoids the time consuming and expensive sequencing and synthesis steps, with computation carried out by DNA strand displacement. We demonstrate the flexibility of our approach by developing schemes for massively parallel binary counting and elementary cellular automaton Rule 110 computation.

Keywords: DNA storage · DNA computing · Parallel computing · Strand displacement

1 Introduction

Studies have espoused DNA as an incredibly dense (up to 455 exabytes per gram) and stable (readable over millenia) digital storage medium [5]. Experiments storing text, images, and movies of hundreds of megabytes have demonstrated the potential scalability of the approach [11]. Importantly, DNA's essential biological role ensures that the technology for manipulating DNA will never succumb to obsolescence.

Typical DNA storage schemes have high information density but do not permit "in-memory" computation: modifying data involves sequencing DNA, classically computing the desired transformation, and synthesizing new DNA. In contrast, strand displacement systems store information in the pattern of reconfiguration of exposed single-stranded regions. This pattern can be directly manipulated through toehold exchange and other molecular primitives as a form of information processing [25]. However, strand displacement is incompatible with traditional DNA storage schemes.

Here we combine DNA storage with massively parallel computation on the data stored using strand displacement. In our proposed scheme, which we call SIMD||DNA (Single Instruction Multiple Data DNA), a multi-stranded DNA complex acts as a single register storing a (binary) string. Although all the

© Springer Nature Switzerland AG 2019
C. Thachuk and Y. Liu (Eds.): DNA 25, LNCS 11648, pp. 219–235, 2019.
https://doi.org/10.1007/978-3-030-26807-7_12

complexes share the same sequence, different information is encoded in each complex in the pattern of nicks and exposed single-stranded regions. There are as many independent registers as the number of molecules of the multi-stranded complexes, each capable of storing and manipulating a different string. This allows information in different registers to be modified at the same time, utilizing the parallelism granted by molecular computation.

Our method of storing information in DNA is motivated by recent developments in DNA storage employing topological modifications of DNA to encode data. DNA storage based on programmable nicking on native DNA (forming strand breaks at desired locations) permits high throughput methods of writing information into registers [20]. To enable subsequent read-out, recently developed methods [9] could potentially read information encoded in nicks and single-stranded gaps in double stranded DNA in a high throughput manner. Reading out specific bits of registers could also be achieved with fluorescence based methods. Note that compared with storing data in the DNA sequence itself, encoding data in nicks sacrifices data density but reduces the cost of large-scale synthesis of DNA [20]. Here we show that it also enables greater flexibility of in-memory computation.

To do parallel in-memory computation on our DNA registers, we employ single instruction, multiple data (SIMD)[1] programs. An overview of a program's implementation is given in Fig. 1. Each instruction of a program corresponds to the addition of a set of DNA strands to the solution. The added strands undergo toehold-mediated strand displacement with strands bound to the register, changing the data. The long "bottom" strands of these registers are attached to magnetic beads, allowing sequential elution operations. After the strands displaced from the registers are eluted, subsequent instructions can be performed. Note that the same instruction is applied to all registers in solution in parallel (since they share sequence space), but the effect of that instruction can be different depending on the pattern of nicks and exposed regions of the given register.

We show that our DNA data processing scheme is capable of parallel, in-memory computation, eliminating the need for sequencing and synthesizing new DNA on each data update. Note that instruction strands are synthesized independently of the data stored in the registers, so that executing an instruction does not require reading the data. We should also note the doubly-parallel nature of SIMD∥DNA programs: instructions act on all registers in parallel, and instruction strands can act on multiple sites within a register in parallel.

---

[1] Single instruction, multiple data (SIMD) is one of the four classifications in Flynn's taxonomy [7]. The taxonomy captures computer architecture designs and their parallelism. The four classifications are the four choices of combining single instruction (SI) or multiple instruction (MI) with single data (SD) or multiple data (MD). SI versus MI captures the number of processors/instructions modifying the data at a given time. SD versus MD captures the number of data registers being modified at a given time, each of which can store different information. Our scheme falls under SIMD, since many registers, each with different data, are affected by the same instruction.

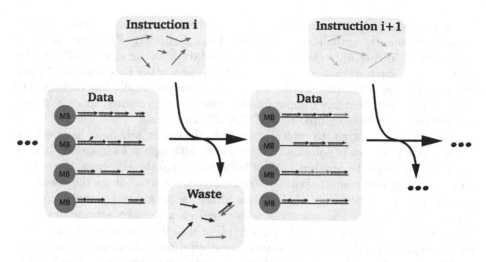

Fig. 1. Each DNA register is a multi-stranded complex. Different information is encoded in the pattern of nicks and exposed single-stranded regions in the register. Registers are attached to magnetic beads (MB). To perform each instruction, first a set of instruction strands is added to the solution and reacts with all the registers in parallel. Then waste species (unreacted instruction strands and displaced reaction products) are washed away by elution.

Our programs require a small number of unique domains (subsequences of nucleotides which act as functional units), independent of the register length. A common assumption for correctness in strand displacement systems is that domains are *orthogonal*, meaning two domains which are not fully complementary do not bind. In experiments, enforcing this assumption requires specialized sequence design. Further, for any fixed domain length, the space of orthogonal domains is limited, restricting the scalability of the system. SIMD||DNA encodes information in the pattern of nicks and exposed domains. This allows our programs to require only a constant set of orthogonal domains to be used (five for one program and six for the other), simplifying the sequence design problem for experimental implementation. In addition, the instruction strands for one instruction can share sequences, resulting in a reduced cost of strand synthesis.

In this paper, we show two SIMD||DNA programs. One of the programs implements binary counting. Starting from arbitrary initial counts stored in different registers, each computation step increments all the registers in parallel. Binary counting allows one SIMD||DNA program to move data through a number of states exponential in the size of the register. We consider this a requirement of any useful data storage/computation suite: if instead not all configurations of the register were reachable from some initial configuration via some program, then the useful density of the storage would be reduced.

In addition to binary counting, we also give a program which simulates elementary cellular automaton (CA) Rule 110.[2] Critically, Rule 110 has been shown to be Turing universal [6], so this simulation shows that SIMD||DNA's in-memory computation model is as powerful as any other space-bounded computing technique. In other words, our space-bounded simulation of Rule 110 immediately gives that any computable function—if the required space is known beforehand—can be computed by a SIMD||DNA program.

We note the contrast to typical strand displacement schemes that perform a single computation in solution. For example, although a logic circuit [13,18] computation might involve hundreds of billions of separate molecules, the redundancy does not help computationally. Such schemes seem to not use the massively parallel power of chemistry [1]. Previous ideas for performing parallel computation with strand displacement cascades relied on a complex scheme involving 4-way branch migration on DNA origami [14] or information processing in different spatial locations [17]. Turing universal computation with strand displacement could not handle multiple independent computations running in parallel [12], leaving the extension to parallel computation as the major open question.

Note that SIMD||DNA is non-autonomous since each instruction requires manual strand addition and elution. In this regard it is similar to early studies of parallel DNA computing machines. Dating back to the 1990s, Adleman experimentally demonstrated solving instances of NP-complete problems using DNA [1], which encouraged other DNA computing models. Many models rely on enzymes to introduce covalent modification on DNA [2,3,8,15], which increases experimental complexity. Other enzyme-free models such as the sticker model [16] encode information in the pattern of exposed domains, similar to our scheme. However, the sticker model requires a number of orthogonal domain types that scales with the amount of data. In addition, these domains require well-tuned binding affinities to allow a melting procedure which selectively dissociates some strands but not others. In contrast, our programs only require a constant number of unique domains for any register length. Instead of computation through controlled hybridization and melting, strand displacement is a more versatile mechanism to achieve modification of information, potentially making parallel molecular computation more feasible.[3]

2   SIMD||DNA

Here we propose the general scheme. First we will explain the notations we use in this paper. We use the domain level abstraction for DNA strands. Consecutive

---

[2] In [24] an enumeration of all possible rules for elementary CA is given. Rule 110 refers to that enumeration.

[3] In a sense, we realize an extension of the sticker model envisioned by [4]: "Recent research suggests that DNA 'strand invasion' might provide a means for the specific removal of stickers from library strands. This could give rise to library strands that act as very powerful read-write memories. Further investigation of this possibility seems worthwhile.".

nucleotides that act as a functional unit are called a domain. Complementary domains are represented by a star (∗). The length of the domains is chosen so that: (1) each domain can initiate strand displacement (can act as a toehold), (2) strands bound by a single domain readily dissociate, and (3) strands bound by two or more domains cannot dissociate.[4] We call an exposed (unbound) domain a *toehold*.

2.1   Encoding Data

Data is stored in multi-stranded complexes (Fig. 1), each called a register. A register contains one long strand, called the *bottom strand* and multiple short strands, called *top strands*, bound to the bottom strand. Each bottom strand is partitioned into sets of consecutive domains called *cells*. Each cell contains the same number of domains. Depending on the configuration of the top strands bound (e.g., their lengths, or the presence or absence of toeholds), cells encode information. In this work we use a binary encoding, with each cell representing one bit.

See Sect. 4.1 for a discussion of potential experimental methods of preparing the initial registers.

2.2   Instructions

An *instruction* is a set of strands. To apply an instruction to the data, these strands are added to the solution at high concentration. Adding these strands can lead to three different types of reactions on the registers. Figure 2a explains the figure notation used to describe instructions throughout the paper, and Fig. 2b gives examples of the three types of reactions. They are:

Attachment: This reaction preserves all the strands originally bound to the register and attaches new strands. An instruction strand can attach to registers if it binds strongly enough (by two or more domains). Note that the attachment of an instruction strand can lead to a partial displacement of a pre-existing strand on the register.

Displacement: This reaction introduces new strands to the register and detaches some pre-existing strands. Upon binding to a toehold on the register, the instruction strand displaces pre-existing strands through 3-way branch migration.[5] Toehold exchange reactions are favored towards displacement by the instruction strand since they are added at high concentration. Two instruction strands can also cooperatively displace strands on the register.

Detachment: This reaction detaches pre-existing strands without introducing new strands to the registers. An instruction strand that is complementary to

---

[4] Given these properties, in practice one could choose the domain length to be from 5 to 7 nucleotides at room temperature.

[5] Although other more complicated strand displacement mechanisms (e.g. 4-way, remote toehold, associative toehold strand displacement) could provide extra power in this architecture, they usually sacrifice the speed and increase the design complexity, so we do not include them in this work.

a pre-existing strand with an open overhang can use the overhang as a toehold and pull the strand off the register. Throughout this paper, a dashed instruction strand indicates the domains in the instruction strand are complementary to other vertically aligned domains.

When an instruction strand displaces a top strand, we assume the waste top strand does not interact further within the system (the instruction strands are present in high concentration while the waste is in low concentration). After the reactions complete, the waste is removed via elution. We assume washing removes all species without a magnetic bead. Lastly, we assume there is no *leak*—displacement of a strand without prior binding of a toehold. We discuss the possibility and extent of errors caused by relaxing these assumptions in Sect. 4.2.

In general, two reactions can be applicable but mutually exclusive. Then two (or more) resulting register states may be possible after adding the instruction strands. The instructions used in this paper do not have this issue. This point is related to *deterministic* versus *nondeterministic* algorithms, and is discussed further in Sect. 4.5.

2.3   Programs

We consider sequences of instructions, called *programs*. We design programs for functions $f : \{0,1\}^n \rightarrow \{0,1\}^n$ so that, given a register encoding any $s = \{0,1\}^n$, after applying all instructions in the program sequentially as in Fig. 1, the resulting register encodes $f(s)$.

3   Programs for Binary Counting and Rule 110

Here we give our two programs: binary counting and simulation of elementary cellular automaton Rule 110. We first present the Rule 110 simulation, as the program is simpler to explain than binary counting.

3.1   Cellular Automaton Rule 110

An elementary one-dimensional cellular automaton consists of an infinite set of cells $\{\ldots, c_{-1}, c_0, c_1, \ldots\}$. Each cell is in one of two states, 0 or 1. Each cell changes state in each timestep depending on its left and right neighbor's states. Rule 110 is defined as follows: the state of a cell at time $t+1$, denoted $c_i(t+1)$, is $f(c_{i-1}(t), c_i(t), c_{i+1}(t))$, where f is the following:

$$f(0,0,0) = 0 \qquad\qquad f(1,0,0) = 0$$
$$f(0,0,1) = 1 \qquad\qquad f(1,0,1) = 1$$
$$f(0,1,0) = 1 \qquad\qquad f(1,1,0) = 1$$
$$f(0,1,1) = 1 \qquad\qquad f(1,1,1) = 0$$

Note that a simple two-rule characterization of f is as follows: 0 updates to 1 if and only if the state to its right is a 1, and 1 updates to 0 if and only if both

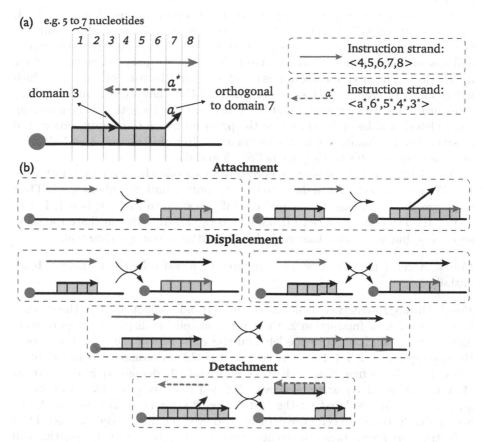

Fig. 2. (a) The notation used to describe instructions. Domains are represented by square boxes. We indicate complementarity of instruction strands to register domains by vertical alignment. If a domain label is given explicitly, such as a and a^* in this figure, the domain is orthogonal to the other vertically aligned domains. A strand can be described by listing the constituent domains in a bracket $<>$ from 5'-end to 3'-end. Strands with solid lines are complementary to the corresponding domains in the bottom strand. Strands with dashed lines are complementary to the corresponding domains in the top strand. The blue dot represents the magnetic bead. (b) The three instruction reactions. Attachment: instruction strands attach to the register without releasing any other strands. Displacement: instruction strands attach to the register and displace pre-existing strands on the register. Toehold-mediated strand displacement (left), toehold-exchange strand displacement (right), and cooperative strand displacement (bottom) mechanisms are allowed. Detachment: instruction strands fully complementary to previously bound top strands pull strands off the register. (Color figure online)

neighbors are 1. This characterization is useful for proving correctness of the program.

The instructions implementing one timestep evolution are shown in Fig. 3. Each state-0 cell is fully covered by two strands, one of length three and one of length two. Each state-1 cell is partially covered by a length-five top strand and has an open toehold at the leftmost domain. The program consists of six instructions. The program first marks the string "01" (Instruction 1)—here, the 0 will change to 1 later. Then it erases the internal 1's in any string of at least three consecutive 1's (Instructions 2 and 3). These are the 1's with two neighboring 1's, which should be updated to 0, so the program fills in the empty cells with 0 (Instruction 4). Finally it removes the markers from Instruction 1 and changes previously marked 0's to 1's (Instructions 5 and 6).

We claim that this program enforces the two-rule characterization of Rule 110. We first argue that 1 updates to 0 if and only if both neighbors are 1. Then we argue that 0 updates to 1 if and only if the state to its right is a 1. Let i_k denote the kth domain on cell i (from left to right). All cells can share the same sequences, but we assume that each domain within a cell is orthogonal.

Claim. A cell i initially in state 1 updates to a 0 if cells $i + 1$ and $i - 1$ are initially 1.

Proof. During Instruction 1, the instruction strands cannot displace the strands in state-1 cells. In Instruction 2, the strand on cell i is displaced cooperatively only if the toeholds on both the left and the right side of the strand are open. By assumption, cell $i + 1$ is a 1, so the toehold immediately to the right of cell i, $(i + 1)_1$, is free. Since cell $i - 1$ is in state 1, domain i_1 is not covered after Instruction 1 (i_1 would be covered if cell $i - 1$ were 0). Thus the strand on cell i can be displaced by the instruction 2 strands. In Instruction 3, the instruction 2 strands in cell i are detached, so every domain in cell i is free. Then in Instruction 4 we attach the strands corresponding to a state 0, updating cell i to 0. Instructions 5 and 6 do not introduce any instruction reaction on cell i, so cell i remains in state 0. □

Claim. A cell i initially in state 1 stays in state 1 if either cell $i + 1$ or $i - 1$ is initially 0.

Proof. During Instruction 1, the instruction strands cannot displace the strands in state-1 cells. In Instruction 2, the strand on state-1 cells is displaced cooperatively only if the toeholds on both the left and the right side of the strand are open. By assumption that the left or right cell is a 0, the toeholds required for this, i_1 or $(i + 1)_1$, will be covered: First consider that cell $i - 1$ is a 0. Then in Instruction 1, the instruction strand displaces one strand at cell $i - 1$ and covers the toehold i_1. On the other hand, if cell $i + 1$ is 0, then domain $(i + 1)_1$ is covered since strands for the 0 state cover all the domains in that cell. So if either neighbor state is 0, Instruction 2 does not displace the strand on cell i. Then note that Instructions 3 and 4 do not introduce any instruction reaction at cell i. The instruction 5 strands detach the instruction 1 strands if cell $i - 1$

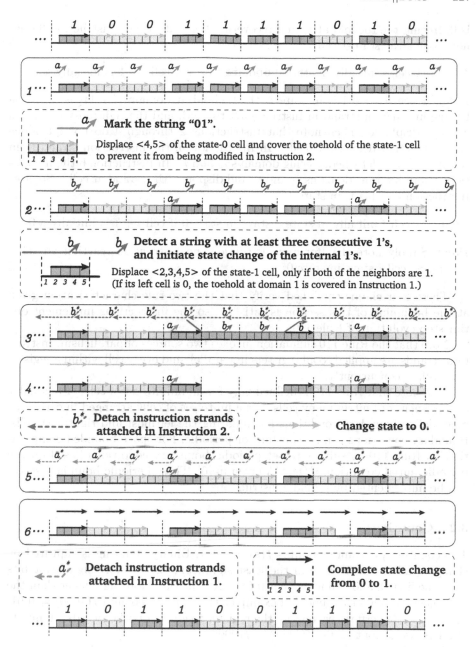

Fig. 3. The program implementing one timestep of Rule 110 shown on an example register. The top register shows the initial state of each cell. After 6 instructions, the register updates to the state shown at the bottom. Strand colors have three information categories: state 1 (dark blue), state 0 (light blue), intermediates (other colors). Solid boxes show the instruction strands and the state of the register before the strands are applied. Dashed boxes explain the logical meaning of the instructions. The overhang domains a and b are orthogonal to their vertically aligned domains. (Color figure online)

is 0, freeing the toehold at i_1 and recovering the state-1 cell. Instruction 6 does not change the state-1 cell. $\square$

Claim. A cell i initially in state 0 updates to a 1 if cell $i + 1$ is initially a 1.

Proof. Since cell $i + 1$ is in state 1, the toehold at domain $(i + 1)_1$ is available for the instruction strand in Instruction 1 to bind, and the rightmost strand on cell i is displaced. Then note that Instructions 2 through 4 do not introduce any instruction reaction at cell i. In Instruction 5, the instruction strand from Instruction 1 is detached, freeing domains i_4 and i_5. In Instruction 6 the instruction strand binds at domains i_4 and i_5 and displaces the strand at cell i. So after Instruction 6, cell i is in state 1. $\square$

Claim. A cell i initially in state 0 stays in state 0 if cell $i + 1$ is initially a 0.

Proof. Simply note that for any instruction, no instruction reaction on cell i occurs. So cell i stays in state 0. $\square$

These four claims combined verify that the two-rule characterization given at the beginning of this section is satisfied, so the instructions implement one timestep evolution of Rule 110.

Note that the Rule 110 simulation invokes two sources of parallelism. Instruction strands are applied to all registers in parallel, and every cell within a register can update concurrently.

Also note that Rule 110 is defined only for an infinite set of cells or a circular arrangement of finitely many cells. For a finite set of cells arranged linearly, one must define boundary conditions for updating the leftmost and rightmost cells. Boundary conditions can be constant or periodic. For space-bounded computation by Rule 110, it suffices to set periodic boundary conditions based on the periodic initial condition of the CA given in [6]. These periodic boundary states can be implemented by periodic instructions.

3.2   Counting

The counting program computes $f(s) = s + 1$. Binary counting is captured by changing all the 1s to 0 from the least significant bit to more significant bits until the first 0, and changing that 0 to 1. All the bits more significant than the rightmost 0 remain the same. For example, $f(1011) = 1100$, and $f(1000) = 1001$. In the case of overflow, we rewrite the register to all 0s. In other words, on inputs of all 1s, we output all 0s: $f(1111) = f(0000)$.

The full program is in Fig. 4. Each state-0 cell is fully covered by two strands, with one covering the first three domains and the other one covering the last two domains. Each state-1 cell is fully covered by two strands, with one covering the first two domains and the other one covering the last three domains. One extra domain is included to the right of the rightmost cell which is used to initiate displacement. The program contains seven instructions. It erases all the 1's in between the rightmost cell and the rightmost state-0 cell at Instructions

1 and 2, and changes those cells to 0 at Instructions 4 and 5. It marks the rightmost state-0 cell at Instruction 3, and change the marked state-0 cell to state 1 at Instructions 6 and 7.

To prove correctness, we first argue that all the 1's from the least significant bit to the rightmost 0 update to 0. Then we argue that rightmost 0 updates to

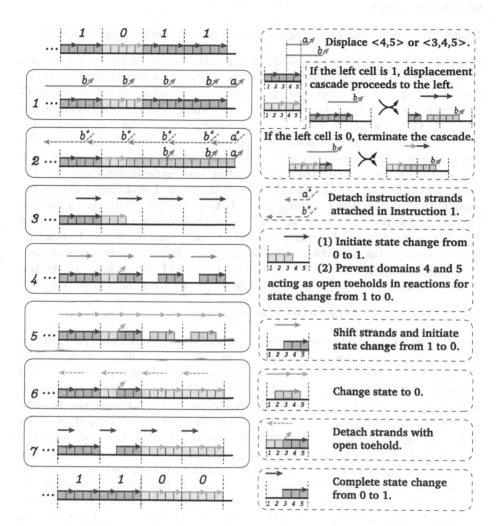

Fig. 4. The program implementing addition by 1 of a binary string on an example register. The top register shows the initial state of each cell. After 7 instructions, the register updates to the state shown at the bottom. Strand colors have three information categories: state 1 (purple), state 0 (pink), intermediates (other colors). Solid boxes show the instruction strands and the state of the register before the strands are applied. Dashed boxes explain the logical meaning of the instructions. The overhang domains a and b are orthogonal to their vertically aligned domains. (Color figure online)

1. Assume the bit string has length n and the least significant bit is at cell n and the rightmost 0 is at cell m ($m \leqslant n$). As in the Rule 110 simulation proof, we let j_k denote the kth domain on cell j (from left to right). All cells can share the same sequences, but we assume that each domain within a cell is orthogonal. Additionally, the extra domain to the right of the rightmost cell is orthogonal to all other domains.

Claim. All state 1 cells to the right of the rightmost 0 cell change to a 0.

Proof. Instruction 1 initiates a series of sequential reactions from the least significant bit n to the rightmost 0. First the instruction strand with overhang domain a displaces the strand covering domains n_4 and n_5. If the least significant bit is 1 ($m < n$), the domain n_3 becomes unbound after this displacement reaction. Then the domain n_3 serves as an open toehold to initiate another displacement reaction with the instruction strand with overhang domain b. Similar displacement reactions proceed until cell m. By assumption, cell m is a state-0 cell, so the domain m_3 will not be open after the displacement step, thus the displacement cascade stops. Then the strands added in Instruction 2 detach the strands from Instruction 1, leaving the cells from the $(m+1)$th bit to the nth bit free. In Instruction 3, every applied instruction strand from cell $m+1$ to n attaches to the register. Instruction 4 shifts those strands added in Instruction 3 one domain to the left, which opens toeholds for the cooperative displacement in Instruction 5. After those cells change to state-0 in Instruction 5, the strands added in Instruction 6 and 7 do not change them, so they remain in state 0.   □

Claim. The rightmost state 0 cell changes to a 1.

Proof. Instruction 1 initiates a series of sequential reactions from the least significant bit to the rightmost 0 at cell m. The domain m_3 will not be open after the instruction strand displaces the strand covering domains m_4 and m_5 and no more strand displacement cascade can proceed to the left. Then the strands added in Instruction 2 detach the strands from Instruction 1, leaving the domains m_4 and m_5 free. The strands added in Instruction 3 serve as two purposes: (1) They correspond to one of the strands representing state 1, thus they help cell m to transition to state 1 and they partially displace the strand at domain m_3. (2) They serve as a place holder by binding at domains m_4 and m_5 to prevent cell m from being modified in Instructions 4 and 5. Instruction 6 detaches the strand originally bound from domain m_1 to m_3, leaving the domains m_1 and m_2 open. In Instruction 7, the instruction strand attaches to the register at domain m_1 and m_2, which completes the state changing from 0 to 1.   □

Claim. The cells to the left of the rightmost state 0 cell stay the same.

Proof. Note that no open toeholds are exposed at cells to the left of cell m, and the displacement cascade does not pass to the left of cell m, thus no changes are made to the states of those cells.   □

4   Discussion and Future Work

4.1   Data Preparation

If we do not try to reuse domain sequences, the registers could be prepared by topologically modifying naturally occurring DNA at desired locations through nicking enzymes[6] [20]. If the distance between two nicks is short (for example the length of one domain), the strand in between will spontaneously dissociate, forming a toehold. After the registers with different information are prepared separately and attached to magnetic beads, they are mixed into one solution.

If we reuse domains between cells, the initial preparation of registers requires different techniques. For example, all registers can be initialized to 0 in separate test tubes, and then separate programs executed which move the registers to the desired initial state.

4.2   Experimental Feasibility and Error Handling

Toehold-mediated strand displacement and elution through magnetic beads are well-established techniques, which supports the feasibility of experimental implementation of SIMD||DNA. Other than attaching registers to magnetic beads, registers can also be affixed to the surface of a microfluidic chip. Further, since the instruction strands are added at high concentration and we do not rely on slow mechanisms such as 4-way branch migration, each instruction should finish quickly. However, strand displacement systems can be error prone, and our constructions make several assumptions, the violation of which could lead to various errors.

The first assumption is that waste products from reactions between the instruction strands and registers do not react further with the system. Registers and instruction strands should be allowed to react for a short amount of time before elution such that the high concentration instruction strands interact with the registers, but the low concentration waste products do not. Violating this assumption can cause undesired displacements to occur, leading to possible error in the computation. Interestingly, we conjecture that, besides the reverse of the intended reaction (in the case of toehold exchange), the waste products and registers cannot react in the two programs given here, and therefore our programs are robust to this type of error.

The next assumption is that of a perfect washing procedure where only species with magnetic beads remain after elution. Imperfect washing can result in previous instruction strands reacting with registers undergoing subsequent instructions. In practice, the remains of imperfect washing would appear in low concentration so as to have a low probability of affecting the system.

The final assumption is that there is no leak (only toehold-mediated displacements occur). The registers contain nicks where strands could fray and undesired

---

[6] For example, Cas9 nickase or restriction enzyme *Pf*Ago, uses an RNA or DNA strand as a guide and can nick at a desired location.

toeholds could open, resulting in strands being mistakenly displaced or incorrect strands binding. Our programs are not robust to leak, raising the question of whether leakless design principles [21–23] can be imposed on the constructions. Leak could also limit the longevity of information stored in our scheme: (toehold-less) four-way branch migration can result in bit exchange errors between different registers.

It remains to be seen whether freezing or other means of stabilizing the DNA complexes suffices to ensure long term storage of information encoded in nicked registers.

In addition to preventing errors at the experimental level, it remains open to address errors at the "software level" by employing error correction codes in the data and employing error correction schemes in the instructions.

4.3   Data Density

Unlike storing data in the DNA sequence itself, which has a data density of 2 bits per nucleotide, our schemes sacrifice data density. In our schemes, a bit is encoded in a cell, which contains 5 domains. If a domain is composed of 6 consecutive nucleotides, it gives a data density of 0.033 (1/30) bit per nucleotide. It is not obvious that the current construction with 5 domains per cell achieves the highest possible data density for these programs. In practice, there is a tradeoff between the strand binding stability and data density. Here we assume that the minimal number of domains required for two strands to stably bind is two, however in practice the binding strength is affected by experimental buffer (salt concentration) and temperature. Given different experimental conditions, it may be necessary to increase the number of domains in a cell, which could reduce the data density further. However, one gram of DNA can still encode exabytes of information. In principle, data density may also be increased by using different encoding schemes, such as allowing overhangs on the top strands to encode information.

4.4   Uniform Versus Non-uniform Instructions

We can identify instructions as *uniform* or *non-uniform*. *Uniform* instructions have the property that the same type of instruction strands are added to every cell, as is the case in our programs. *Non-uniform* instructions allow strands to be added to particular cells and not others (e.g., add strands to every seventh cell, or cells 42 and 71). The difference in computational power between uniform and non-uniform instructions remains open, and non-uniform instructions could reduce the number of instructions required for some programs. However, non-uniform instructions could require each cell to be orthogonal in sequence. In contrast, uniform instructions allow every cell to consist of the same sequence, requiring only the domains within the cells to be orthogonal. Sharing the sequences between the cells reduces the number of different instruction strands that need to be synthesized.

4.5   Determinism and Nondeterminism

Our programs are designed with *deterministic* instructions: given one state of the register, after adding the instruction strands, the register changes to one specific state. Deterministic instructions make it easy to design, predict, reason about, and compose the programs. In contrast to deterministic instructions, one could also construct *nondeterministic* instructions by introducing nondeterminism to the updates of the cells. For example, consider an empty cell with domains $\langle 3^*, 2^*, 1^* \rangle$, and add instruction strands $\langle 1, 2 \rangle$ and $\langle 2, 3 \rangle$. Either the first or second strand can bind, but since they displace each other, only one will remain after elution. The probability of which strand remains depends on its relative concentration. In principle, applying nondeterministic instructions allows for implementation of randomized algorithms and simulation of nondeterministic computing machines.

4.6   Running Time

The running time of a program depends on two factors: running time per instruction and the number of instructions. The running time per instruction depends on whether the instruction updates the cells through *parallel* or *sequential* reactions. In general, instructions are capable of acting on each cell within each register in parallel. Yet, Instruction 1 of the binary counting program does not have this source of parallelism. A first reaction (displacement) must occur on the rightmost cell prior to a second reaction occurring on the second cell, which must occur prior to a third reaction on the third cell, and so on. Thus, this instruction with sequential reactions loses the speedup given by independent instruction reactions occurring in parallel on each cell within a register. Besides the running time per instruction, the larger the number of instructions per program, the more complex is the experimental procedure. This motivates studying the smallest number of instructions required to achieve a computational task.

4.7   Universal Computation

Our registers as proposed are restricted to a finite number of cells. So although Rule 110 on an infinite arrangement of cells can simulate an infinite-tape Turing machine, our scheme is only capable of space-bounded computation. To claim that a system is capable of universal computation, it is required that the data tape—in our case, the number of cells—can be extended as needed as computation proceeds. Since our program consists of uniform instructions, domain orthogonality is only required within a cell. Therefore, in principle, the register can be extended indefinitely during computation without exhausting the space of orthogonal domains. The register's length could perhaps be extended by merging bottom strands with top strand "connectors".

4.8 Space-Efficient Computation

Although Rule 110 is Turing universal, computing functions through simulation of a Turing machine by Rule 110 does not make use of the full power of SIMD||DNA. First of all, while simulation of a Turing machine by Rule 110 was shown to be time-efficient [10], it is not space-efficient. Precisely, simulating a Turing machine on an input which takes T time and $S \leq T$ space requires $p(T)$ time and $p(T)$ space (where $p(T)$ is some polynomial in T). However, Turing machines can be simulated time- and space-efficiently by one-dimensional CA if the automaton is allowed more than two states [19]. Simulating larger classes of CA is a promising approach to space-efficient computation in this model, since our Rule 110 simulation suggests that CA are naturally simulated by SIMD||DNA programs.

4.9 Equalizing Encodings

Our two programs use different schemes for encoding binary information in a register. Using some universal encoding would allow applying different consecutive computations to the same registers. Alternatively, we could design programs to inter-convert between different encodings. The reason for suggesting this alternative is that unlike classical machines acting on bits, in SIMD||DNA the way a bit is encoded affects how it can be changed by instruction reactions. For example, in the binary counting program, the encoding ensures that no toeholds except for the rightmost domain are open on the register, which is used to argue correctness. Alternatively, in the Rule 110 program, toeholds must be available throughout the register to achieve the parallel cell updates required by CA. Therefore having one encoding which implements these two different functions seems difficult.

Acknowledgements. The authors were supported by NSF grants CCF-1618895 and CCF-1652824, and DARPA grant W911NF-18-2-0032. We thank Marc Riedel for suggesting the analogy to Single Instruction, Multiple Data Computers. We thank Marc Riedel and Olgica Milenkovic for discussions.

References

1. Adleman, L.M.: Molecular computation of solutions to combinatorial problems. Science **266**(5187), 1021–1024 (1994)
2. Beaver, D.: A universal molecular computer. DNA Based Comput. **27**, 29–36 (1995)
3. Boneh, D., Dunworth, C., Lipton, R.J., Sgall, J.: On the computational power of DNA. Discret. Appl. Math. **71**(1–3), 79–94 (1996)
4. Braich, R.S., Chelyapov, N., Johnson, C., Rothemund, P.W.K., Adleman, L.: Solution of a 20-variable 3-SAT problem on a DNA computer. Science **296**(5567), 499–502 (2002)
5. Church, G.M., Gao, Y., Kosuri, S.: Next-generation digital information storage in DNA. Science **337**(6102), 1628 (2012)

6. Cook, M.: Universality in elementary cellular automata. Complex Syst. **15**(1), 1–40 (2004)

7. Flynn, M.J.: Some computer organizations and their effectiveness. IEEE Trans. Comput. **21**(9), 948–960 (1972)

8. Freund, R., Kari, L., Păun, G.: DNA computing based on splicing: the existence of universal computers. Theory of Comput. Syst. **32**(1), 69–112 (1999)

9. Liu, K., et al.: Detecting topological variations of DNA at single-molecule level. Nat. Commun. **10**(1), 3 (2019)

10. Neary, T., Woods, D.: P-completeness of cellular automaton rule 110. In: Bugliesi, M., Preneel, B., Sassone, V., Wegener, I. (eds.) ICALP 2006. LNCS, vol. 4051, pp. 132–143. Springer, Heidelberg (2006). https://doi.org/10.1007/11786986_13

11. Organick, L., et al.: Random access in large-scale DNA data storage. Nat. Biotechnol. **36**(3), 242–248 (2018)

12. Qian, L., Soloveichik, D., Winfree, E.: Efficient turing-universal computation with DNA polymers. In: Sakakibara, Y., Mi, Y. (eds.) DNA 2010. LNCS, vol. 6518, pp. 123–140. Springer, Heidelberg (2011). https://doi.org/10.1007/978-3-642-18305-8_12

13. Qian, L., Winfree, E.: Scaling up digital circuit computation with DNA strand displacement cascades. Science **332**(6034), 1196–1201 (2011)

14. Qian, L., Winfree, E.: Parallel and scalable computation and spatial dynamics with DNA-based chemical reaction networks on a surface. In: Murata, S., Kobayashi, S. (eds.) DNA 2014. LNCS, vol. 8727, pp. 114–131. Springer, Cham (2014). https://doi.org/10.1007/978-3-319-11295-4_8

15. Rothemund, P.W.K.: A DNA and restriction enzyme implementation of turing machines. DNA Based Comput. **27**, 75–119 (1995)

16. Roweis, S., et al.: A sticker-based model for DNA computation. J. Comput. Biol. **5**(4), 615–629 (1998)

17. Scalise, D., Schulman, R.: Emulating cellular automata in chemical reaction-diffusion networks. Nat. Comput. **15**(2), 197–214 (2016)

18. Seelig, G., Soloveichik, D., Zhang, Y., Winfree, E.: Enzyme-free nucleic acid logic circuits. Science **314**(5805), 1585–1588 (2006)

19. Smith III, A.R.: Simple computation-universal cellular spaces. J. ACM **18**(3), 339–353 (1971)

20. Tabatabaei, S.K., et al.: DNA punch cards: encoding data on native DNA sequences via topological modifications. bioRxiv, 10.1101/672394

21. Thachuk, C., Winfree, E., Soloveichik, D.: Leakless DNA strand displacement systems. In: Phillips, A., Yin, P. (eds.) DNA 2015. LNCS, vol. 9211, pp. 133–153. Springer, Cham (2015). https://doi.org/10.1007/978-3-319-21999-8_9

22. Wang, B., Thachuk, C., Ellington, A.D., Soloveichik, D.: The design space of strand displacement cascades with toehold-size clamps. In: Brijder, R., Qian, L. (eds.) DNA 2017. LNCS, vol. 10467, pp. 64–81. Springer, Cham (2017). https://doi.org/10.1007/978-3-319-66799-7_5

23. Wang, B., Thachuk, C., Ellington, A.D., Winfree, E., Soloveichik, D.: Effective design principles for leakless strand displacement systems. Proc. Nat. Acad. Sci. **115**(52), E12182–E12191 (2018)

24. Wolfram, S.: Statistical mechanics of cellular automata. Rev. Mod. Phys. **55**, 601–644 (1983)

25. Zhang, D.Y., Seelig, G.: Dynamic DNA nanotechnology using strand-displacement reactions. Nat. Chem. **3**(2), 103 (2011)

Author Index

Printed in the United States
By Bookmasters